Riverbank Filtration: An Evaluation of RBF Hydrology and Impacts on Yield

Awwa Research Foundation

About the Awwa Research Foundation

The Awwa Research Foundation (AwwaRF) is a member-supported, international, nonprofit organization that sponsors research to enable water utilities, public health agencies, and other professionals to provide safe and affordable drinking water to consumers.

The Foundation's mission is to advance the science of water to improve the quality of life. To achieve this mission, the Foundation sponsors studies on all aspects of drinking water, including supply and resources, treatment, monitoring and analysis, distribution, management, and health effects. Funding for research is provided primarily by subscription payments from approximately 1,000 utilities, consulting firms, and manufacturers in North America and abroad. Additional funding comes from collaborative partnerships with other national and international organizations, allowing for resources to be leveraged, expertise to be shared, and broad-based knowledge to be developed and disseminated. Government funding serves as a third source of research dollars.

From its headquarters in Denver, Colorado, the Foundation's staff directs and supports the efforts of more than 800 volunteers who serve on the board of trustees and various committees. These volunteers represent many facets of the water industry, and contribute their expertise to select and monitor research studies that benefit the entire drinking water community.

The results of research are disseminated through a number of channels, including reports, the Web site, conferences, and periodicals.

For subscribers, the Foundation serves as a cooperative program in which water suppliers unite to pool their resources. By applying Foundation research findings, these water suppliers can save substantial costs and stay on the leading edge of drinking water science and technology. Since its inception, AwwaRF has supplied the water community with more than $300 million in applied research.

More information about the Foundation and how to become a subscriber is available on the Web at **www.awwarf.org.**

Riverbank Filtration: An Evaluation of RBF Hydrology and Impacts on Yield

Prepared by:
Stephen A. Hubbs
University of Louisville, W.S. Speed 101, Louisville, KY 40292
and
Kay Ball and **Tiffany Caldwell**
Louisville Water Company, 550 South 3rd Street, Louisville, KY 40202-1409

Jointly sponsored by:
Awwa Research Foundation
6666 West Quincy Avenue, Denver, CO 80235-3098
and
Louisville Water Company
550 South Third Street
Louisville, KY 40202-1509

Published by:

Awwa Research Foundation

American Water Works Association

IWA Publishing

DISCLAIMER

This study was jointly funded by the Awwa Research Foundation (AwwaRF) and Louisville Water Company (LWC). AwwaRF and LWC assume no responsibility for the content of the research study reported in this publication or for the opinions or statements of fact expressed in the report. The mention of trade names for commercial products does not represent or imply the approval or endorsement of either AwwaRF or LWC. This report is presented solely for informational purposes.

CONTENTS

TABLES

FIGURES

FOREWORD

The Awwa Research Foundation is a nonprofit corporation that is dedicated to the implementation of a research effort to help utilities respond to regulatory requirements through a process of consultation with subscribers and drinking water professionals. Under the umbrella of a Strategic Research Plan, the Research Advisory Council prioritizes the suggested projects based upon current and future needs, applicability, and past work; the recommendations are forwarded to the Board of Trustees for final selection. The foundation also sponsors research projects through the unsolicited proposal process; the Collaborative Research, Research Applications, and Tailored Collaboration programs and various joint research efforts with organizations such as the U.S. Environmental Protection Agency, the U.S. Bureau of Reclamation, and the Association of California Water Agencies.

This publication is a result of the one of these sponsored studies, and it is hoped that its findings will be applied in communities throughout the world. The following report serves not only as a means of communicating the results of the water industry's centralized research program but also as a tool to enlist the further support of nonmember utilities and individuals.

Projects are managed closely from their inception to the final report by the foundation's staff and large cadre of volunteers who willingly contribute their time and expertise. The foundation serves a planning and management function and awards contracts to other institutions such as water utilities, universities, and engineering firms. The funding of this research effort comes primarily from the Subscription Program, through which water utilities subscribe to the research program and make an annual payment proportionate to the volume of water they deliver and consultants and manufacturers subscribe based on their annual billings. The program offers a cost-effective and fair method for funding research in the public interest.

A broad spectrum of water supply issues is addressed by the foundation's research agenda: resources, treatment, and management. The ultimate purpose of the coordinated effort is to assist water suppliers to provide the highest possible quality of water economically and reliably. The true benefits are realized when the results are implemented at the utility level. The foundation's trustees are pleased to offer this publication as a contribution toward that end.

Walter J. Bishop
Chair, Board of Trustees
Awwa Research Foundation

Robert C. Renner, P.E.
Executive Director
Awwa Research Foundation

ACKNOWLEDGMENTS

The research described in this document was completed through the cooperation of the Louisville Water Company (LWC), the University of Louisville (U of L), and the United States Geological Survey in a research project jointly funded by the Louisville Water Company, the United States Geological Survey, and the Awwa Research Foundation.

We would like to express our appreciation to the Project Advisory Committee for their oversight, ideas, and thorough review of submittals:

M. Robin Collins, University of New Hampshire
William Gollnitz, Greater Cincinnati Water Works
Jay Jasperse, Sonoma County Water Agency

The contributions of several individuals were instrumental in completing this research. Dr. D. Joseph Hagerty of U of L provided continuous guidance, and Dr. Tom Rockaway managed the overall project at the University. Kay Ball of LWC coordinated activities between U of L, LWC, and AwwaRF. We would like to thank Mike Unthank of the U.S. Geological Survey at the Louisville office for his contribution to the research effort. Also, we like to thank the water quality and operations staff at LWC who provided assistance in the study, and Larry Bryant, who retrieved valuable data from the SCADA system.

The efforts of numerous other individuals also contributed to the completion of this study. We would like to acknowledge the utilities and individuals that provided the data that formed the basis for this study:

Louisville Water Company, Louisville, KY
Greater Cincinnati Water Works, Cincinnati, OH
Sonoma County Water Agency, Sonoma Co., CA
Kansas City Board of Public Utilities, Kansas City, MO
Kansas City Water Services, Kansas City, MO
Des Moines Water Work, Des Moines, IA
City of Lincoln, Lincoln, NE
Collector Wells, International
Jim Constantz, Menlo Park, CA
Matthew Damos, Sonoma County, CA
Don Seymour, Sonoma County, CA
William Gollnitz, Cincinnati, OH
Bruce Whitteberry, Cincinnati, OH
Jürgen Schubert, Dusseldorf, Germany
Jordi Martín-Alonso, Barcelona, Spain
Bernhard Wett, Innsbruck, Austria
Durward Johnson, Kansas City, KS
Eric Lee, Lincoln, NE
Ted Corrigan, Des Moines, IA

EXECUTIVE SUMMARY

A utility considering riverbank filtration (RBF) is typically faced with a few basic questions about the technology:

- How many and what type of wells are needed?
- What spacing is required between the wells and how much riverbank will be needed?
- What will it look like aesthetically?
- How much will it cost?

This report provides input to the first two of these questions, addressing the basic hydrology of RBF systems. This report focuses on the interaction between the river, the riverbed, the aquifer, and the well along with the associated parameters that influence how much water will flow from a given system at a given site.

BACKGROUND

Many utilities have been turning to RBF as an alternative to raw river water as they seek to improve the quality of source water used for water supply. The effectiveness of RBF for removal of microbial contaminants and natural organic matter (NOM) was demonstrated through a research project co-funded by the Louisville Water Company (LWC) and the Awwa Research Foundation (AwwaRF) at a radial collector well in Louisville, Kentucky in 2002 (Wang, Hubbs, and Song 2002). During the 2-year monitoring of the RBF well at Louisville, the well experienced a decline in production from 22 mgd to 17 mgd, which was attributed to possible clogging of the riverbed.

The decline in production and lack of understanding of factors that affect clogging in RBF systems prompted LWC to initiate research into the clogging process and the factors that impact the long-term yield from RBF systems. The research was co-funded by AwwaRF, LWC, and the United States Geological Survey (USGS) and performed in cooperation with these organizations and the University of Louisville Civil and Environmental Engineering Department.

The primary parameter evaluated in this study was specific capacity (pump discharge per unit drawdown). Site and water quality characteristics thought to impact specific capacity were also evaluated, including stream velocity and slope, riverbed shear stress, riverbed media composition, aquifer and riverbed hydraulic conductivity, and river water quality and temperature. This study was designed to compile the available data on these parameters, and to advance the understanding of the riverbed/aquifer interactions.

RESEARCH OBJECTIVES

The overall objective of the research was to investigate the clogging process in RBF systems and identify factors significant to yield. Specific objectives of this research included:

1. evaluate commonly available stream and aquifer characteristics from many RBF systems and distinguish the impact of these characteristics on yield,
2. collect extensive data from the RBF system at Louisville and apply the information to advance the understanding of riverbed clogging, and
3. provide recommendations regarding the design of RBF systems in order to minimize riverbed clogging and long-term yield decay at other RBF sites.

APPROACH

Data on aquifer characteristics, riverbed characteristics, site layout, and river hydrology, and water quality were collected from 11 RBF systems in the United States and Europe. These data were compiled into tables by category for comparison and analysis. The data set obtained was only moderately complete and proved inadequate for detailed regression analysis. Instead the data were analyzed using simple comparisons to the operational performance of each study site. Adequate data were available from four sites to demonstrate the significance of water temperature to well yield.

Data from the radial collector well in Louisville were analyzed to characterize the conditions of the riverbed and monitor the change in riverbed conditions with time. This work included simultaneous monitoring of production rate, well level, river level, well temperature, and river temperature, as well as periodic monitoring of pressure levels under the riverbed and flux rates across the riverbed.

CONCLUSIONS

Data collected from 11 RBF systems in the United States and Europe and the site-specific data from Louisville lead to the following conclusions:

1) RBF has been implemented successfully in a variety of complex conditions in the United States and Europe. The data compiled in this study provide a background against which future RBF sites can be compared.
2) The riverbed/aquifer interaction is critical in the recharge and long-term capacity of RBF systems. This interaction should be considered in the design of RBF systems.
3) Riverbed clogging in RBF systems can result from mechanical, biological, and geochemical processes.
4) The impact of unsaturated conditions under the riverbed on riverbed hydraulic conductivity and well specific capacity is dramatic, resulting in a decrease in specific capacity and the extension of the point of recharge further from the well.
5) The impact of temperature on yield from RBF systems is dramatic because of the wide range of water temperature experienced at RBF sites. A strong correlation was developed between specific capacity and temperature at individual sites, as shown by data from four RBF systems. Specific capacity of an RBF system can more than double as temperatures in the aquifer and river fluctuate from annual lows to annual highs.
6) Data from the 11 sites were inadequate to quantify the effects of other parameters such as river discharge and river water quality on the capacity of RBF systems. This uncertainty was likely due to limitation in the number of sites studied.
7) Data from Louisville showed that specific capacity can decrease significantly in early stages of system development.

RECOMMENDATIONS

The following recommendations are made from this project:

1) Design Considerations.
   a) RBF should be considered for water supply in areas with transmissive alluvial aquifers that are hydraulically connected to a river source.
   b) The hydrogeologic investigation should be combined with an evaluation of the hydrology of the river, the water quality of the river and ground water, and the aerial extent and depth of the aquifer at the site.
   c) A pump test should be performed capable of differentiating between aquifer conductivity and river recharge.
   d) The impact of seasonal variations in temperature on the capacity of a proposed RBF system should be considered in the design and initial hydrogeologic testing of the site.
   e) Data on riverbed sediment and aquifer media composition (particle size analysis) should be collected from available literature and augmented with samples collected at the site.
2) Systems should be designed with the realization that a decline in capacity in the first few years is possible. However, measures can be taken to reduce the expected decline in yield by:
   a) Sizing the facility to minimize the impact of unsaturated conditions under the riverbed.
   b) Designing an extraction system that distributes infiltration stresses evenly along the riverbed.
   c) Locating the facility to maximize the regenerative impacts of streambed scouring, with preference to locations with movable streambeds, avoiding areas of streambed armoring and locations where very fine sediment is deposited.
3) RBF facilities should maintain records allowing specific capacity to be tracked as a function of well and river temperature, and time.

FUTURE RESEARCH

1) While it is known that scour is important to the renovation of riverbed surfaces in RBF systems, the intensity and depth of beneficial scouring is site specific, with little data available in the literature. Additional research into the depth of clogging and required depth of scour to yield beneficial riverbed renewal is warranted.
2) The compilation of basic operational data from RBF facilities will lend increasing insight into the significance of riverbed and aquifer characteristics in RBF design. Specifically, temperature data between the river and the well discharge reflect a lag in peaks, a shift in averages, and a magnitude difference between maximum-to-minimum values. These three parameters can yield valuable information regarding riverbed and aquifer hydraulic conductance and groundwater contribution to the well discharge. These operational data should be recorded and analyzed for trends.
3) This research has indicated that relatively large hydraulic forces can be exerted on the riverbed when the media below becomes unsaturated. This phenomenon should be further explored at the laboratory level, evaluating the hydraulic conductivity and shear resistance before and after exposure to unsaturated conditions.

CHAPTER 1
INTRODUCTION

STATEMENT OF PROBLEM

Water quality improvements from riverbank filtration (RBF) have made RBF a desirable alternative to the traditional use of river water for water supply. Many utilities are turning to RBF because of savings in treatment costs compared to the treatment of raw river water. Determining the maximum sustainable yield from a given site is a challenge, however, as methods of assessing the long-term recharge capacity from the riverbed of the adjacent stream have not been adequately developed. Traditionally, yield has been predicted using an estimate of aquifer transmissivity and riverbed conductance from a short-term pump test at the site. This prediction method fails to fully account for influences of site geometry, river hydraulics, sediment transport, and source water quality on long-term riverbed recharge and well yield. Although short-term pump tests may be adequate for predictions of initial capacity, experience has shown that RBF systems can exhibit a trend of declining specific capacity in the first few years of operation.

The clogging of the riverbed is believed to be one of the major contributors to the long-term decay of well yield. A better understanding of riverbed clogging and other factors that impact yield is needed to allow for more effective management of existing systems and better design of future systems. This project was designed to advance the understanding of clogging in RBF systems by compiling data from various RBF sites and by extensively evaluating the RBF site at Louisville, Kentucky.

BASICS OF RIVERBANK FILTRATION

Traditional sources of public water supply include surface water and groundwater. Surface water is often readily available, but requires extensive treatment due to contamination from both natural and anthropogenic sources. Groundwater is typically a higher quality source than surface water, and usually requires less treatment that surface waters. Groundwater is naturally renewed through processes such as infiltration of precipitation and flow from surface water sources into aquifers. But in many areas, groundwater is not available in quantities capable of meeting the water demand.

RBF has been an acknowledged treatment process in Europe for over 100 years. RBF could be described as a hybrid source between surface water and groundwater. A typical RBF system consists of one or a series of wells adjacent to a river in an alluvial aquifer that is hydraulically connected to that river. Pumping from the wells induces flow from the river through the bank toward the well. This induced flow acts to continuously replenish the water pumped from the well. As the water from the river flows through the river bottom and aquifer, it is naturally filtered by the bank and aquifer materials. The removal process is thought to be similar to that of slow sand filtration, a combination of both filtration and biological degradation.

RBF has proven effective in greatly reducing the level of contaminants in the river sourcewater, including biodegradable organics, pathogens, and particles. Previous work at Louisville (Wang, Hubbs, and Song 2002) provided an extensive evaluation of the water quality improvements typical from RBF facilities. The hydrology of the RBF system at Louisville presented in this report is intended to compliment the previous water quality work at this site.

Yield-limiting Factors in RBF

The elements of an RBF system impacting long-term sustainable capacity of a wellfield can be divided into three yield-limiting factors: riverbed conductance; aquifer transmissivity; and wellscreen dynamics (Figure 1.1). Each of these factors needs to be considered when designing RBF systems, along with the total driving head available within the aquifer.

The hydraulic capacity of a stream to recharge a wellfield is influenced by a number of factors including stream flow, riverbed scouring and sediment transport, width and configuration of the riverbottom, the area of exposure of the aquifer to the riverbed, the conductivity of the riverbed, water temperature, and the biological processes taking place within the riverbed. These factors are illustrated in Figure 1.1 as "riverbottom", and are not typically considered when designing RBF systems. The hydraulic conductivity of a riverbed can be expected to change with time, both seasonally with riverbed scour and deposition, and over the course of years if the regenerative forces of scour are inadequate to restore the hydraulic capacity of the riverbottom. These factors will be discussed in depth in Chapters 3 and 4.

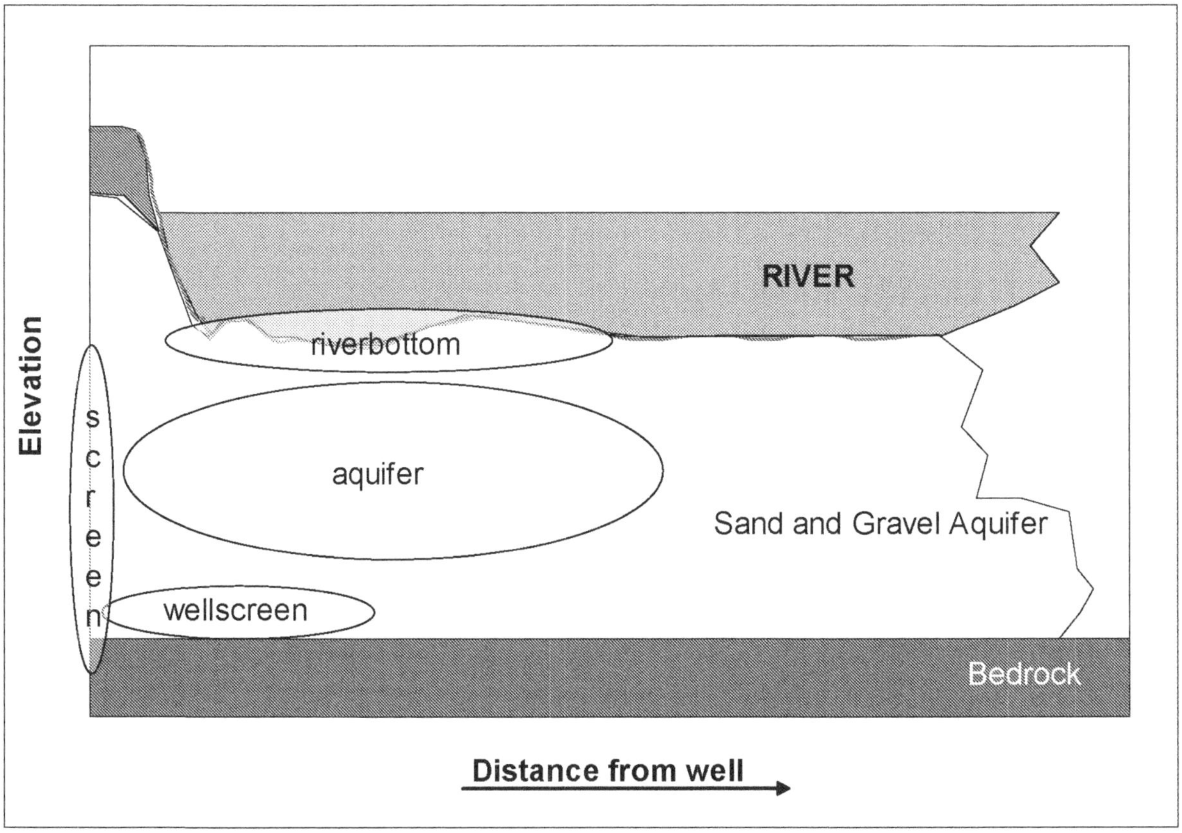

Figure 1.1 Factors limiting yield in RBF systems

The capacity of the aquifer to transport water from the river to the wellscreen is largely determined by the size and homogeneity of the sands and gravels making up the bulk of the aquifer. Medium sand to fine gravel is optimum, sorted at the riverbed so that very little fine sand and silt penetrate the more productive larger media. Many Midwestern streams that were formed during glacial periods have deep ancient riverbeds with thick deposits of well sorted gravel towards the bottom of the aquifer. These deep gravel deposits can act as an underdrain, allowing recharge from a relatively large area. Horizontal collector wells are particularly well suited to extract water from these coarser layers. The presence of boulders in these layers, however, can limit the ability to install laterals, and thus caution should be taken to avoid alluvial deposits containing boulders when designing horizontal collector systems.

A properly designed and conducted pumping test will yield adequate indications of aquifer hydraulic conductivity and riverbed connectivity. Because of the desired interaction between the river and the aquifer, pump tests must be interpreted carefully to extract and separate information on aquifer conductivity and river recharge. In RBF systems, anisotropy (the ratio of horizontal conductivity to vertical conductivity) can be important in systems because the predominant flowpaths can be both vertical and horizontal. In most instances, the hydraulic conductivity of the aquifer (excluding the riverbed and wellscreen areas) does not change appreciably with time.

Wellscreen head loss can be influenced by hydraulic entrance velocity, chemical deposition, biological growth, and wellscreen development during construction. The entrance velocity can be affected to a moderate extent by the diameter of the wellscreen selected for vertical wells; however, the length of wellscreen is limited by the depth of the aquifer. Horizontal wells can be constructed with adequate length of screen to greatly reduce the impact of entrance velocity on the capacity of the well. Chemical deposition and biological growth on and near the screen can cause deterioration in well yield with time, and can be the limiting factor for a particular site. With any construction technique, the importance of properly developing the well to remove fine particles from the well drilling process and in the aquifer is critical, and if improperly executed can result in decreased system yield.

In vertical wellscreen systems, the limited amount of wetted screen and high velocities around the wellscreen are important variables affecting the long-term capacity of a well system. Advances in radial well design, however, have allowed for up to 500 m of horizontal wellscreen to be developed in formations where vertical wells can accommodate only 20 m of wellscreen. In many formations, this change has eliminated head loss around the wellscreen as the limiting factor in sustainable well capacity.

The final hydraulic consideration in RBF systems is the driving head between the river source and the well. As the well is operated, the water level in the well or caisson provides the driving head (or energy) between the river source and the well. This energy is consumed by the processes described above: water infiltration into the riverbed, conveyance through the aquifer, and acceleration immediately around and through the wellscreen. The greater the discharge rate from the well, the greater the headloss in each of these processes, and the greater driving head required to pull water through the system. Shallower aquifers have less available head, and thus typically have less total capacity, given similar riverbed, aquifer, and wellscreen characteristics.

Recent investigations have indicated that riverbed conductance can be the capacity-limiting factor in high-capacity RBF systems. However, the impact of riverbed conductance and its change with time on long-term sustainable yield has not been thoroughly evaluated. It has been observed at Louisville that riverbed conductance varies as a function of time, which is likely the result of riverbed clogging. This clogging can be caused by mechanical particle impingement, biological growth, or geochemical reactions within the aquifer/riverbed interface. All three of these processes can be impacted by dynamic and static hydraulic forces. However, the impact of these clogging processes on the long-term specific capacity of a wellfield is poorly understood.

RESEARCH OBJECTIVES

The overall objectives of the research were to investigate the clogging processes and identify factors significant to yield. This report presents results of an investigation of riverbed conditions at Louisville and a compilation of data from several sites to identify site characteristics important to yield. Specific objectives of this research included:

1. evaluate commonly available stream and aquifer characteristics from many RBF systems and distinguish the impact of these characteristics on yield,
2. collect extensive data from the RBF system in Louisville at the B.E. Payne Water Treatment Plant and apply the information to advance the understanding of riverbed clogging, and
3. provide recommendations regarding the design of RBF systems and minimize riverbed clogging and long-term yield decay.

APPROACH

The approach used in this project was to assess RBF hydrology broadly using commonly available data from various RBF sites, and to collect detailed site-specific data on one site to help interpret the other data. Site-specific studies were performed at the radial collector well in Louisville to characterize the conditions of the riverbed and monitor the change in riverbed conditions with time. These studies included simultaneous monitoring of production rate, well level, river level, well temperature and river temperature, as well as periodic monitoring of pressure levels under the riverbed, riverbed conductance, and riverbed conditions in and around the infiltration area of the riverbed.

A separate research focus was an investigation of factors significant to yield using data from several research sites. This research was stochastic in nature. Data were collected from 11 RBF systems in the United States and Europe relative to the aquifer characteristics, riverbed characteristics, site layout, and river hydraulics, hydrology, and water quality for each site studied. These data were compiled into tables by category. The data set obtained was only moderately complete and the sparse nature of the information prevented detailed regression analyses or analyses of variances from being performed. Instead the data were analyzed using simple comparisons of the data collected for each factor and the operational performance of each study site. Adequate data were available from four sites to demonstrate the significance of temperature in the river and aquifer to yield.

CHAPTER 2
METHODS AND MATERIALS

DATA COLLECTION FROM PARTICIPATING SITES

Data were compiled from 11 RBF systems and analyzed to identify factors significant to yield. Study sites were selected based on data availability and the level of support from operating utilities. Site characteristics included in the study were selected based on the impact to system yield. Data were collected through Internet research, literature searches, site visits, and email correspondence. The data were compiled into tables by category and analyzed. An analysis was performed to evaluate the impact of temperature on RBF system performance using data from four study sites.

Site Selection

Sites selected for this study were chosen on the basis of the amount and quality of data available, the ease of obtaining data, and the level of support from the utilities operating the systems. The selected sites included six from the United States and five from Europe. Most of the United States sites were located in the central part of the country where conditions are favorable for RBF. Sites in the United States included wellfields along:

- the Platte River near Ashland, Nebraska,
- the Missouri River in Kansas City, Missouri,
- the Raccoon River in Des Moines, Iowa,
- the Ohio River in Louisville, Kentucky,
- the Great Miami River near Cincinnati, Ohio, and
- the Russian River in Sonoma County, California.

In Europe, sites were selected along:

- the Rhine River at Düsseldorf, Germany,
- the Enns River at Steyr, Austria, and
- the Llobregat River at Cornellà, Spain.

Site locations in the United States and in Europe are shown in Figure 2.1 and Figure 2.2, respectively. An additional site near Albany, New York on the Hudson River was also evaluated. However, the data needed to compare the Albany site to the other sites were not available.

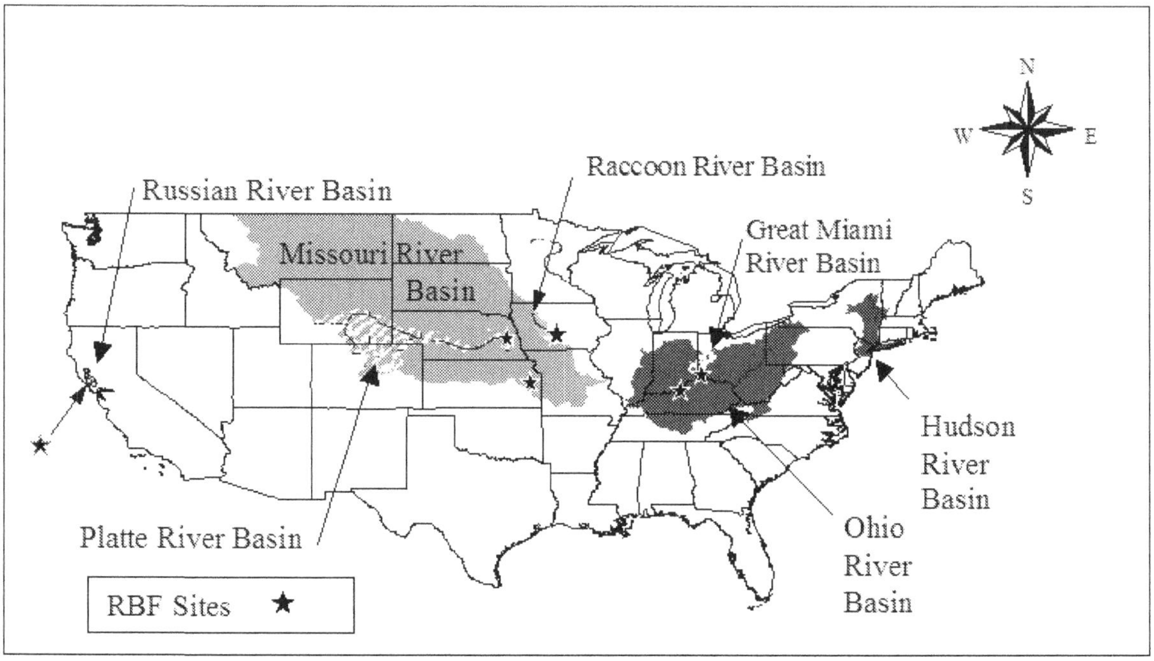

Figure 2.1 Map of sites in the United States (Data from USGS)

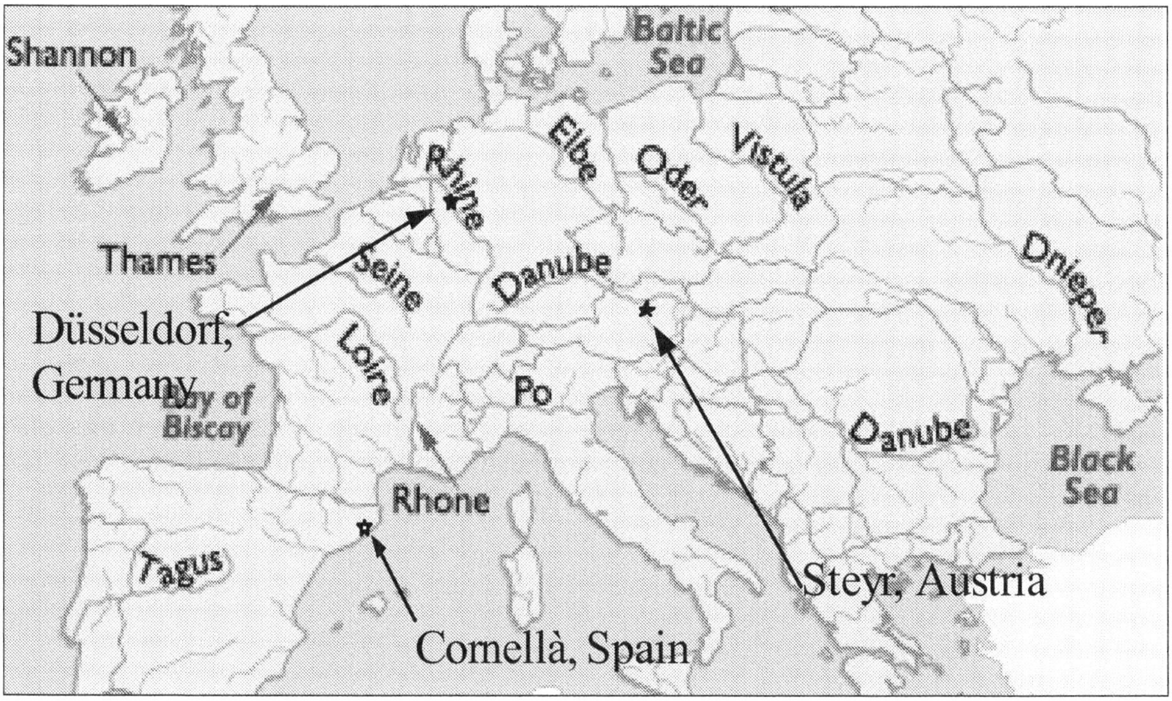

Source: Adapted from Worldatlas.com 2005.

Figure 2.2 Map of sites in Europe

Site Data Overview

The selected sites represent RBF systems with different types of wells along various types of rivers (meandering, straight, and braided rivers). The wells were categorized into two main types: vertical wells and radial collector wells. A third type of well, an inclined horizontal well, was utilized at one site; however, information for this well was not available. Therefore, the inclined horizontal well was excluded from the study and only information from other wells at that site was utilized.

The majority of study sites in Europe utilized vertical wells whereas the majority of sites studied in the United States utilized radial collector wells. Capacities for United States sites ranged from approximately 45,000 m^3/d to 140,000 m^3/d (12 to 37 mgd). Capacities of the European sites ranged from 1,700 m^3/d to 106,000 m^3/d (0.5 to 28 mgd). The average daily production for each site is shown graphically in Figure 2.3 Site information is summarized in Table 2.1.

Site Descriptions

This section includes a short site summary for each location included in the study. The summaries describe the location, the wellfield, and the river.

Ohio River at Louisville, Kentucky

The Louisville Water Company began operating a single radial collector well in Louisville, Kentucky in August 1999 (Wang, Hubbs, and Song 2002). The well produces an average of 64,000 m^3/d (17 mgd). The well is located adjacent to the Ohio River, which acts

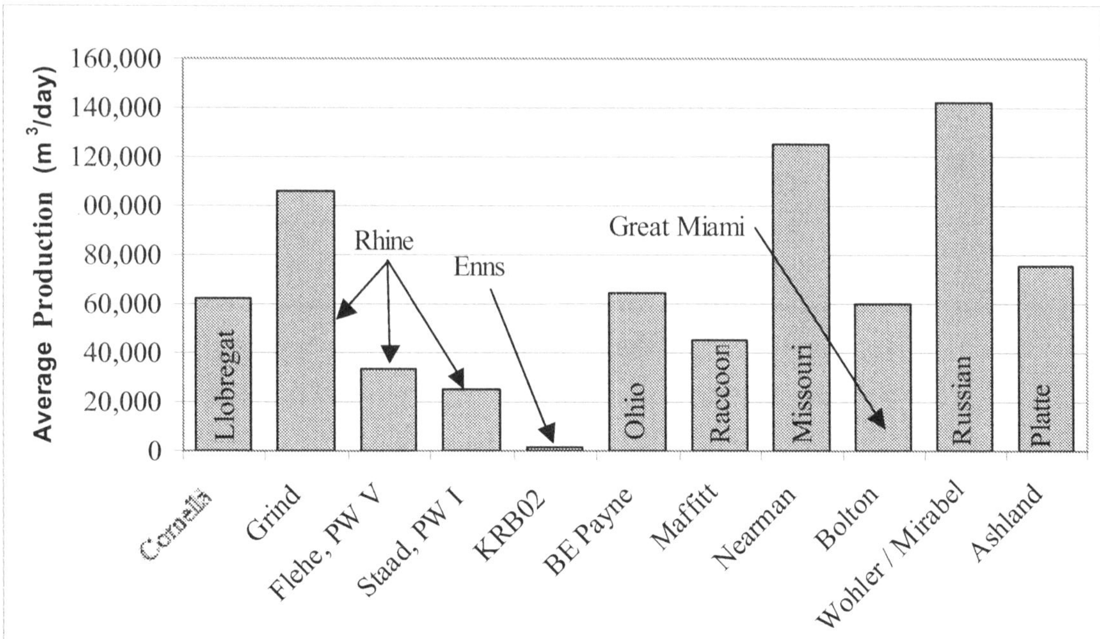

Figure 2.3 Average production by study site

Table 2.1 Sites selected for study

Location	Waterworks	Wellfield	Description	Average Production (m^3/day)
United States				
Great Miami River near Cincinnati, OH	**Greater Cincinnati Water Works**	**Charles M. Bolton**	**10-vertical**	**60,000**
Missouri River in Kansas City, KS	**Kansas City Board of Public Utilities**	**Nearman WTP**	**1-radial**	**120,000**
Ohio River in Louisville, KY	**Louisville Water Company**	**B.E. Payne WTP**	**1-radial**	**64,000**
Platte River near Ashland, NE	**City of Lincoln**	**Ashland**	**2-radial**	**76,000**
Raccoon River in Des Moines, IA	**Des Moines Water Works**	**Maffitt**	**5-radial** *	**45,000** *
Russian River in Sonoma Co., CA	**Sonoma County Water Agency**	**Wohler and Mirabel**	**5-radial**	**140,000**
Europe				
Location	Waterworks	Wellfield	Description	Average Production (m^3/day)
River Enns in Austria	**Ennskraft Company**	**KRB02**	**1-vertical**	**1,700**
Llobregat River in Spain	**Barcelona Water Company**	**Cornellà**	**26-vertical and 7-Recharging Wells**	**62,000**
Rhine River in Germany	**Flehe**	**Flehe, PW V**	**70-vertical**	**33,000**
Rhine River in Germany	**Grind**	**Grind**	**7-radial**	**106,000**
Rhine River in Germany	**Staad**	**Staad, PW I**	**25-vertical**	**25,000**

* The RBF site on the Raccoon River consists of 5 radial wells and 1-large inclined horizontal well. However, the inclined horizontal well was excluded from the study. The production data shown includes production from the 5 radial wells only. The inclined horizontal well produces an additional 5,000 m^3/d on the average

as the source of recharge. The Ohio can be characterized as one of the larger rivers included in the study. It is comparable in size to the Rhine River and is slightly larger than the Missouri River. It is approximately 10 m deep and 600 m wide at the well location. The water surface elevation, or stage, of the Ohio River is affected by numerous navigational locks and dams located along its length. Those dams were constructed to maintain minimum depths of water in navigation channels. The dams do not affect discharge in the river, on average. The mean discharge for the Ohio River at Louisville is approximately 3,300 m^3/s. The Ohio River flows into the Mississippi River approximately 450 km downstream of the well.

Great Miami River at Cincinnati, Ohio

The Greater Cincinnati Water Works operates the Charles M. Bolton Wellfield in Southwestern Ohio near Cincinnati (Sheets, Darner, and Whitteberry 2002). The wellfield consists of 10 vertical wells spaced approximately 250 m apart adjacent to the Great Miami River. These wells draw water from the Great Miami River Buried Valley Aquifer, which is hydraulically connected to the Great Miami River (Gollnitz et al. 2003). The 10 vertical wells are used in rotation and operated to meet demand (Gollnitz, Whitteberry, and Vogt 2004). The wellfield produces an average of 60,000 m^3/d (12 mgd). The river itself is wide and shallow during normal flow. Average discharge is approximately 100 m^3/s. This river is similar in size and shape to the Russian River and Raccoon River. The Great Miami River is a tributary to the Ohio River; it enters the Ohio River approximately 56 km downstream of the study site.

Raccoon River at Des Moines, Iowa

The Des Moines Water Works (DMWW) operates five radial collectors and one large inclined horizontal well adjacent to the Raccoon River in Des Moines, Iowa. According to data provided by the DMWW, the five radial collectors produce an average of about 45,000 m^3/d (12 mgd) combined while the single inclined horizontal well produces approximately 5,000 m^3/d (1.3 mgd). However, the inclined horizontal well was excluded from the study because of inadequate data relative to the layout, operation, and monitoring of that well. All information included in this study for the Raccoon River site is relevant to the five radial collector wells only. The wellfield is approximately 15 km upstream from the junction of the Raccoon and Des Moines Rivers. The Raccoon River is a relatively small river; average discharge is approximately 50 m^3/s. This river is wide and shallow like the Russian River and Great Miami River; it is a principal tributary to the Des Moines River.

Platte River near Ashland, Nebraska

The City of Lincoln operates two radial collector wells on an island in the Platte River in Ashland, Nebraska (Hensen, Obrist, and Prior undated). The two wells are spaced approximately 1,000 m apart and produce a combined average of approximately 76,000 m^3/d (20 mgd) (Lee 2004). Average discharge of the river is approximately 160 m^3/s. The Platte River is a broad shallow stream. It becomes a braided channel during periods of low flow. The Platte is similar in depth to the Raccoon, Russian, and Great Miami Rivers, but its width is approximately 10 times greater.

Russian River in Sonoma County, California

The Sonoma County Water Agency operates two wellfields adjacent to the Russian River at Mirabel and at Wohler Bridge. The Mirabel wellfield is on the inside (convex side) of a bend of the Russian River; and consists of three radial collector wells and 7 vertical wells (Constantz 2006). The Wohler Bridge wellfield is upstream of the Mirabel wellfield and consists of two radial collector wells. The wells are spaced an average of 300 m apart. An additional collector well is under construction, equipped with two 45 cm (18-in) diameter laterals 115 m long and ten 30 cm (12-in) laterals. An inflatable dam is raised downstream of Wohler Bridge during periods of low flow. This provides for increased recharge to the aquifer from the river and allows pumping into higher elevation infiltration basins to further recharge the aquifer (Jasperse and Beach 2000). The annual average production is approximately 140,000 m^3/d (37 mgd) and the average peak day production at this site is approximately 318,000 m^3/d (84 mgd). The Russian River itself is a wide shallow stream during normal flow. Average discharge is approximately 40 m^3/s, but flow is very flashy. Flow from spring through fall is controlled by releases from two reservoirs. The Russian is similar in size and depth to the Raccoon River and Great Miami River. The Russian River becomes an estuary 37 km downstream of the wellfield at Jenner, California and effectively is an arm of the Pacific Ocean from that point downstream.

Missouri River at Kansas City, Kansas

The Kansas City Board of Public Utilities began operating a radial collector well adjacent to the Missouri River in Kansas City, Missouri in August 2000 (National Water Research Institute 2005). According to Collector Wells International, INC (2005), the well is the largest alluvial collector well in the world; it produces an average of 120,000 m^3/d (33 mgd) and is capable of producing up to 170,000 m^3/d (45 mgd). The Missouri River is similar to the Ohio and Rhine Rivers, but slightly smaller. Average discharge is approximately 1500 m^3/s. The river is 4 m deep and 300 m wide under normal flow conditions. The Missouri River flows into the Mississippi River approximately 590 km downstream of the site location. Like the Ohio River, the Missouri River is considered a principal tributary to the Mississippi River.

Llobregat River at Cornellà, Spain.

The Barcelona Water Company (Aigües de Barcelona) operates a field of vertical wells near the Llobregat River in Cornellà, Spain. The wellfield consists of 33 vertical wells, including 26 extracting wells and 7 recharging wells (Martín-Alonso 2004). The wells are spaced approximately 200 m apart. The 26 extracting wells produce an average of 62,000 m^3/d (16 mgd). The Llobregat River is a small river with a width of approximately 25 m, and average discharge of approximately 40 m^3/s (Martín-Alonso 2004). The Llobregat enters the Mediterranean Sea downstream of the wellfield, near Barcelona, Spain.

Many aspects of the Cornellà site distinguish it from other sites. The river near the wellfield is not hydraulically connected to the aquifer (Martín-Alonso 2004). Water infiltrates in a narrow zone of high riverbed permeability several km upstream of the point of extraction, and flows horizontally through the aquifer to the location of the wells. Under natural flow conditions, a layer of silt blocks the infiltration of water and without mechanical intervention the riverbed is fully clogged. Infiltration is enhanced by routine scarification of the riverbed with a bulldozer. During flooding, infiltration occurs all along the riverbed, resulting in significant recharge to the aquifer. The Barcelona Water Company also uses recharge wells to improve recharge to the aquifer.

The setting of the RBF system on the Llobregat River also distinguishes it from other systems. Because of its proximity to the Mediterranean Sea, drawdown of the phreatic surface from water withdrawal causes the intrusion of salt water into the aquifer (Martín-Alonso 2005). This RBF system is described in detail in Martín-Alonso (2006).

Rhine River at Düsseldorf, Germany.

The Grind, Staad, and Flehe Waterworks all operate wellfields adjacent to the Rhine River at Düsseldorf, Germany. Because of their close proximity and similar characteristics, the three wellfields are treated herein as a group. The Rhine River is one of the larger rivers included in this study. It is similar to the Ohio and Missouri Rivers in size and shape, but it is slightly larger than the Missouri River. Average discharge of the Rhine River at Düsseldorf is approximately 2100 m^3/s. The Rhine River flows into the North Sea approximately 420 km downstream of Düsseldorf.

The Grind wellfield is the furthest upstream of the three wellfields. The wellfield consists of seven radial collector wells spaced approximately 500 m apart on the inside (convex side) of a bend on the left descending bank of the Rhine River (Schubert 2004). The seven radial collector wells combined produce an average of 106,000 m^3/d (28 mgd).

The Flehe wellfield begins approximately 5 km downstream from the end of the Grind wellfield on the outside (concave side) of a bend on the right descending bank of the Rhine River (Schubert 2004). The wellfield consists of 70 vertical wells spaced approximately 20 m apart, connected by a siphon pipe. The well gallery produces an average of approximately 33,000 m^3/d (9 mgd).

The Staad wellfield begins approximately 16 km downstream from the end of the Flehe wellfield in a straight section on the right descending bank of the Rhine River (Schubert 2004). The wellfield consists of 25 vertical wells spaced approximately 20 m apart, connected by a siphon pipe. The well gallery produces an average of approximately 25,000 m^3/d (7mgd).

Enns River at Steyr, Austria.

The Ennskraft Company operates a single experimental vertical well on the Enns River near Steyr, Austria (Wett 2006). The well produces an average of 1700 m^3/d (0.4 mgd). Though the Enns River is very steep in its upper reaches, the water surface slope at the study location is minimal because of the backwater effect of a reservoir near the wellfield. The river completely penetrates the aquifer, and the riverbed and aquifer sit on a low permeability flysch (sandstone) layer: thus, recharge from the river to the aquifer occurs only through the riverbank and not the riverbed. The Enns River is approximately 5 m deep and 80 m wide with an average discharge of approximately 160 m^3/s (Wett 2006). The Enns River enters the Danube River approximately 35 km downstream from Steyr, Austria.

Data Collection

Each site included in the study was investigated to determine the values for the factors identified previously as significant to yield. For sites in the United States, much of the required information came from the United States Geologic Survey (USGS 2004), the National Weather Service (NWS 2004), literature searches, and site visits. Data from the USGS were used extensively. For each site, stage measurements, streambed sediment analyses, and water quality data were collected from the USGS website for a gage station near the RBF site. Discharge measurements were used to calculate average discharge, yearly maximum discharge, and yearly minimum discharge. Discharge values and information on river cross-section could be used to estimate average velocities. Stage measurements were used to determine stream elevations along each river. This information was combined with river mile/kilometer information from USGS

Water Data Reports to generate water surface profiles for each river. When river mile/kilometer information was not available from USGS, distances were estimated using river maps. The completed profiles were used to calculate stream surface slopes. Water quality data were used to estimate water quality parameters such as suspended solids concentration at important discharge rates. Streambed sediment data were used to evaluate the D_{10}, D_{50}, and D_{90} sizes[1]. In addition, river geometry and temperature measurements were available for certain sites.

Bankfull and flood discharges were available from the National Weather Service Advanced Hydrologic Prediction Service and United States Army Corps of Engineers (USACE) for many of the sites. Plan and cross-section views of RBF facilities were acquired from literature searches, site visits, and email correspondence. These drawings were used to determine the geometric site conditions, type of wellfield, location relative to river bends, and various other measurements.

Site visits proved valuable for verifying information previously collected and for collecting new data. Sites that were visited included the Russian River in Sonoma County, California, the Great Miami River near Cincinnati, Ohio, the Ohio River in Louisville, Kentucky, and the Llobregat River in Barcelona. During the site visits, observations were made on the riverbed materials where they were visible and on the materials that were being transported by the streams. Information from operating utilities was collected during the site visits and consisted of plan and cross sectional views of the sites, reports from studies performed, and raw operational data.

The information collected through the methods previously described was compiled into a table that summarized the data available for all the sites included in the study. Blank cells were used to represent missing information. The table was sent by email to the utilities that operated the sites, researchers who studied the sites, or other individuals who were otherwise knowledgeable of the RBF facility. Those individuals were asked to verify the information in the table and provide the information needed to complete the table. This process was especially useful for collecting data from European sites because literature and web searches provided little information for those sites. In fact, most of the information for European sites was acquired through email correspondence.

Much of the data acquired consisted of ranges of values. If a range of values was obtained for a particular factor, attempts were made to reduce the range into a representative value for easy comparison to the corresponding values from other sites included in the study. This was done by using available information and choosing the most representative value or by averaging the maximum and minimum values of the range when additional information was not available.

Methods of Estimating Length of Riverbank Impacted

Hydraulic interference between wells in an RBF system occurs as the zone of influence from multiple wells overlap. Proper spacing of wells is an important design consideration, with the goal being to balance maximum yield per length of riverbank with minimal interference between wells. The length of riverbank impacted by an RBF system is difficult to estimate without detailed data from monitoring wells along the river. In this study, information from

[1] Filter media is often described as the weight percentage of material less than a specific sieve screen. A D_{10} value of 1 mm indicates that 10% of the material in a given sample is less than 1 mm.

individuals experienced with the RBF systems was used where available to estimate this length of impact. Otherwise, the length was estimated by observing the layout of the wellfield.

For vertical wells, the length was assumed to be equal to the length of the wellfield plus half the distance between the outer wells and the nearest adjacent well. A single radial collector was assumed to impact 600 m of riverbank unless available information indicated otherwise. This assumption was derived from drawdown data from the Ohio River, which indicated that the single horizontal well at that location impacted approximately 600 m of riverbank. For multiple radial collector wells, the length was estimated using the same procedure as for the vertical wellfield. The length impacted was assumed to be the length of wellfield plus half the distance between the outer wells and the nearest adjacent well. Though these procedures produce a crude estimation of length impacted, the results should be sufficiently accurate for comparison between study sites. Ideally, the lengths would be estimated from locations of negligible drawdown or certain small percent of the total drawdown. Such a procedure would reduce the uncertainty in the estimate of impacted length and improve the accuracy of any analysis.

Summary of Factors

The factors that were identified as potentially impacting yield are summarized by category in Figure 2.4.

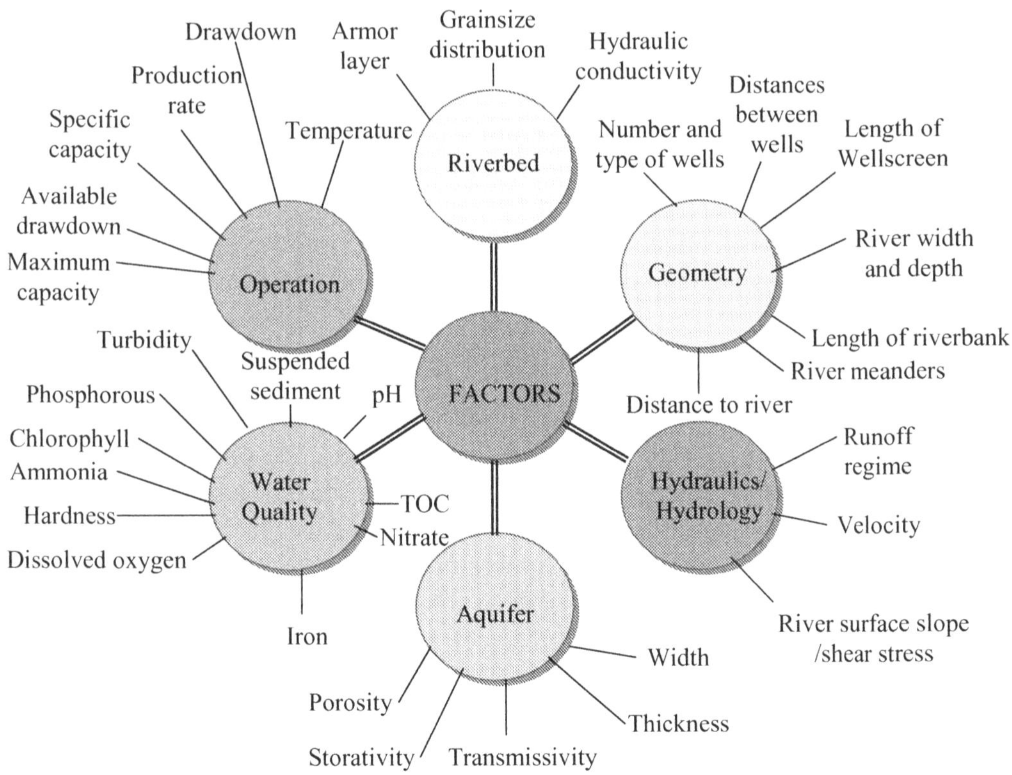

Figure 2.4 Summary of factors identified as potentially significant to yield.

Data Analysis

A regression analysis was performed using time series temperature and specific capacity data from four study sites. The data needed for the analysis included river level, well water level, production rates, and temperature data for the aquifer and/or river. In the case where both well and river temperature data were available, the analysis was run twice; once using the well temperature data, then again using the river temperature data. Only the results of the analysis that resulted in the highest regression coefficient (R^2) were included in this report.

Quality Assurance/Quality Control

Data for this study were collected and compiled from a wide range of sources and compiled into tables for comparative purposes. The tables were sent to the individuals who supplied the information to verify that the information was interpreted and entered into the database correctly. The data were then subjected to a critical review in which the validity of data points from each that were outside expected ranges was verified.

An important aspect to keep in mind when evaluating the data is that the data points were collected from many different sources in the U.S. and Europe. Ideally, the data for all sites included in the study would have been generated by comparable methods and reported in a standard way. However, some data points came from very specific tests while others were estimates by professionals familiar with the site. This procedure introduced some uncertainty into the analysis of the data. The data provided represent the best data available, and are considered adequate for comparison purposes.

Data on surface water hydrology and aquifer characteristics (transmissivity, conductivity, porosity) were taken from sources including USGS publications, groundwater consultants, and utilities. Water quality data were provided by the participating utilities in the U.S. and Europe. Stream sediment data and suspended solids data were taken from USGS literature and data from the utilities. No attempt was made to determine the quality of data from these various sources.

METHODS-LWC SITE

The measurement of aquifer pressure, riverbed shear stress, and riverbed conductivity at the Louisville site are described in the text of this report within the pertinent chapters. Methods used for specific equipment (piezometers) followed the manufacturer's recommendations. Methods used for flux measurements were modifications of techniques described in the literature. River level, well flow, and temperature data were taken from the LWC data acquisition system, with equipment calibrated according to manufacturer's recommendations.

CHAPTER 3
STREAM HYDROLOGY

In RBF systems, stream hydrology determines the composition of the riverbed and the regenerative forces required to keep the riverbed from clogging. The deposition of sediment at the riverbed surface must be balanced by regenerative scouring in order for an RBF system to be sustainable. Streambed scouring can occur at low flows with only a few centimeters of riverbed impacted, or can occur at high flowrates impacting the riverbed to depths of hundreds of centimeters.

Scouring is the result of fluid forces (shear stress) exerted on the riverbed materials. The extent of scouring is determined by the magnitude of the shear stress and the size and density of the riverbed particle. The shear stress is a function of fluid velocity at the streambed, and the viscosity and density of the fluid. The velocity of the fluid at the streambed is a function of stream surface slope and resistance to flow imparted by the streambed. Thus, streambed scouring at a given RBF site can be highly variable, based on localized variations in flow velocity (inside and outside of bends), fluid viscosity (as influenced by temperature and sediment load at the streambed), and riverbed composition (size and density of riverbed material). All of these parameters would be expected to vary with time and location, and determine the sediment in transport at any given time on a streambed.

In this chapter, the basics of stream scour and sediment transport will be discussed, along with an analysis of hydraulic forces imparted on streambed particles by fluid passing over and through the riverbed. Stream hydrographs and streambed slopes are presented, followed by a compilation of stream hydrology data from the participating sites. A brief review of streambed scour resulting from barge traffic is presented at the end of the chapter, and detailed analysis of shear stress estimations at Louisville using slope, velocity, and sediment data are presented in Appendix B.

STREAM SCOUR AND SEDIMENT TRANSPORT BASICS

The science of erosion and sediment transport was highly developed in the 1970's, driven primarily by the concern for scouring around highway bridges and bridge piers and the sedimentation of reservoirs, and more recently by the concern for the effects of sediment deposition on stream aquatic life. The process of erosion has modified the surface of the earth and sorted the sediments that form many of the alluvial aquifers used for water supply. This process is powered by the energy of falling water, with the rate controlled by the composition and resistance of the soils and rocks exposed to the flowing water (Leopold, Wolman and Miller 1992).

The energy potential of a stream can be directly measured by the change of elevation of the stream surface with distance, or the slope of the stream. The energy level in the headwaters of streams is typically high, causing the fine sands and gravels to be carried downstream, leaving large rocks and cobbles in the streambeds. As the energy level decreases with moderate slopes in the mid-reaches of a stream, sands and gravels may be deposited and sorted into highly productive aquifers. If the rates of deposition vary widely with geologic time, distinct zones and layers of media of varying particle size can result in complex aquifers with variable capacity for water supply. As stream slopes decrease near the delta area at the mouth of a stream, finer sands, silts and clays are typically deposited, limiting the conductance between the stream and the adjacent aquifer.

The capacity of a stream for riverbed scour and sediment transport can thus be inferred by considering the slope of the stream through a river profile. Although many of the streams in this study support similarly sized RBF systems, the river hydrology and riverbed scouring characteristics vary greatly among the system sites.

Many streams are impacted by control structures (locks and dams) for flood control, navigation, and power supply. The appropriate slope to use for streambed shear stress and sediment transport is the slope within the pool of the dam, which is often a fraction of the natural stream slope. If one is primarily concerned with annual maximum scouring rates, however, and if these control structures are routinely submerged by flood flows, the average stream slope across several pools can be used to provide a rough estimate of average streambed shear forces under these conditions. The change in elevation of the streambed or top-of-bank can also be used to estimate annual maximum historical stream slope in the absence of any other data.

The sediment transported by streams is determined by the composition and weathering of rocks and soils within the watershed, and the resuspension of riverbed materials from upstream. Periods of high and consistent flow during glacial periods produced many large, well sorted aquifers with good hydraulic connectivity to the streambeds. Materials in transport today may reflect the continued transport of upstream glacial deposits, or newer gravel, sand, silt and clay from upgradient weathering of rocks and soils. Many of the more productive RBF aquifers have a highly transmissive aquifer overlain with a floodplain of more recently deposited silts and clays.

PARTICLE FORCE DIAGRAM

The hydraulic forces imparted onto an individual particle (sand or gravel) on a riverbed may be divided into static forces and dynamic forces. In most evaluations of riverbed shear and sediment transport, the dynamic forces are the most obvious, with the moving water imparting laminar and turbulent forces on the particle. Static forces include particle buoyancy and the overall weight of the water column at a given point. Since equal static forces usually are assumed to be exerted on the riverbed particle from all directions, the weight of the water column is not typically considered in the overall evaluation of forces on the particle.

Forces in Saturated Streambeds Without Induced Infiltration

Julien (1998) described the shear forces acting upon the particles of a riverbed, as presented in Figure 3.1. This force diagram illustrates the stream forces required to impart a shear force adequate to cause movement of the particle on the streambed surface, which occurs when the lift and drag forces overcome the particle weight and adjacent particle resisting forces. This diagram assumes no velocity gradient into or out of the riverbed, as would occur whenever a stream is gaining or losing to the adjacent aquifer.

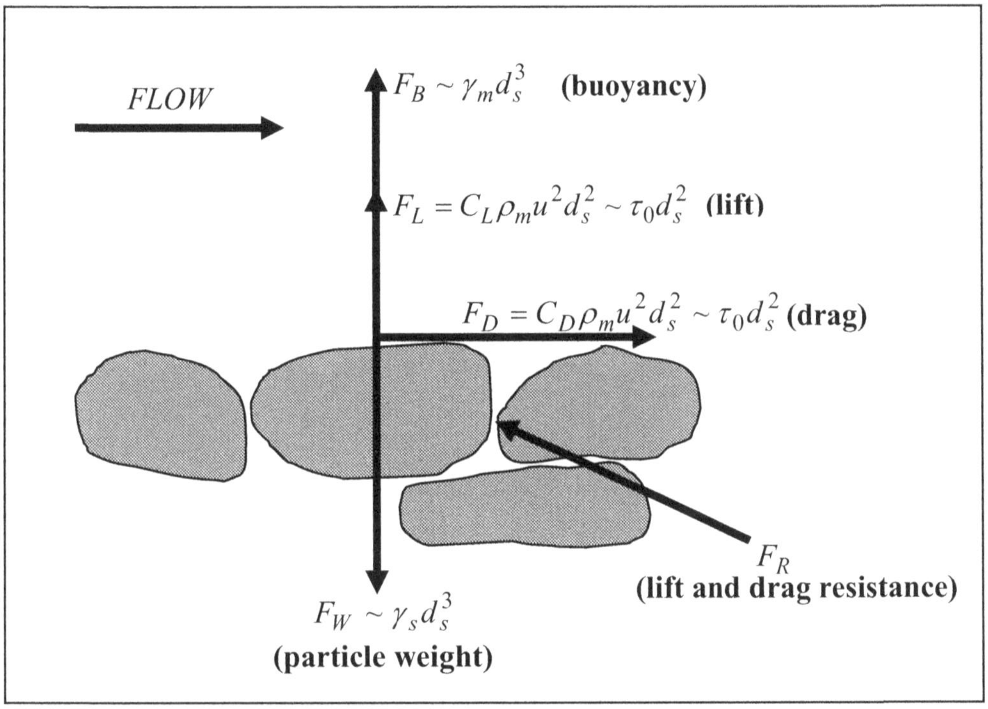

Figure 3.1 Riverbed force diagram without infiltration forces (adapted from Julien, 1998. *Erosion and Sedimentation*. Reprinted with the permission of Cambridge University Press)
F_B is the buoyant force, F_L is the lift force, F_D is the drag force, F_W is the particle weight, and F_R is the resisting force. γ_m is the specific weight of water, γ_s is the specific weight of the sediment particle, d_s is the particle size, d_s^3 represents the particle volume, ρ_m is the fluid density, C_D is a drag coefficient, C_L is the lift coefficient, τ_o is the streambed shear stress, and u is the fluid velocity.

Forces Imparted by Infiltration in RBF Systems

To evaluate the relative magnitude of forces that come into play when a velocity gradient is imparted into the riverbed or riverbank, the fluid viscosity, vertical velocity, and riverbed particle area perpendicular to the flow must be considered. The vertical velocity forces become complicated as the water passes into the aquifer; the approach (Darcy) velocity is a fraction of the interstitial velocity, which is impacted by the porosity of the formation. For a typical sand aquifer with a porosity of 20%, the interstitial velocity within the aquifer is 5 times greater than the Darcy velocity at the approach to the riverbed. The fluid velocity within the aquifer is driven by a hydraulic gradient, and if the aquifer is uniform and not impacted by clogging, the velocity is constant and the head loss increases linearly with distance. This force is shown in Figure 3.2 as the infiltration force F_I.

If the riverbed/aquifer interface is impacted by clogging, additional forces develop from head losses concentrated at the riverbed. If the aquifer under the riverbed becomes unsaturated, the static force of the overlying water column is exerted directly on the riverbed. These forces can be significant and can impact both the hydraulic conductivity and shear resistance of the riverbed. Unsaturated conditions have been indicated in the literature (Constantz 2006) and in this project.

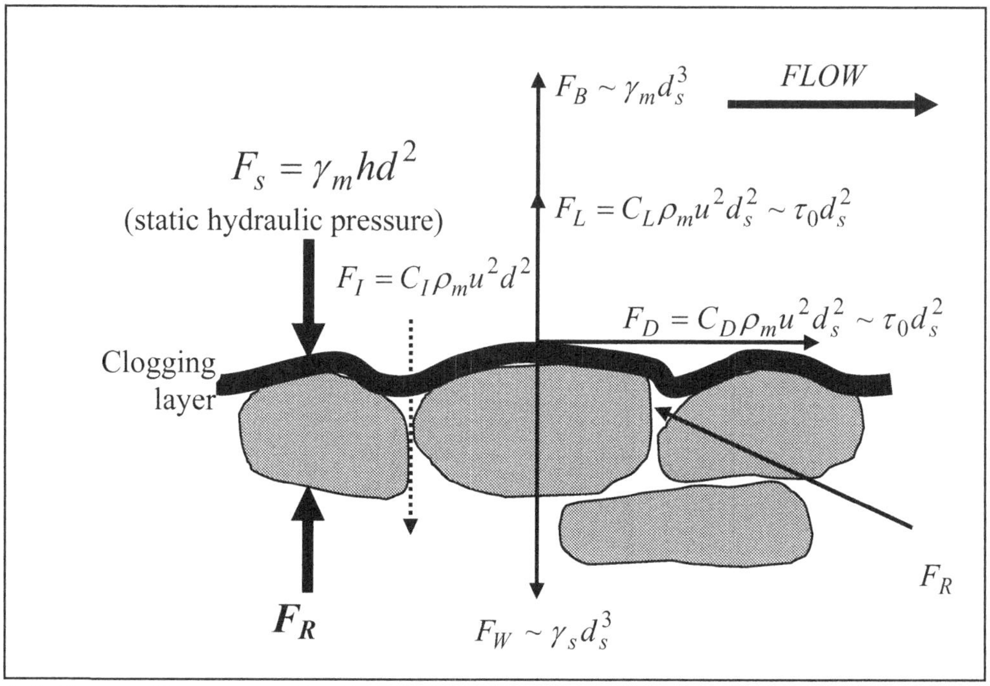

Figure 3.2 Force diagram reflecting unsaturated conditions with static forces shown (adapted from Julien, 1998. *Erosion and Sedimentation*. Reprinted with the permission of Cambridge University Press)
F_I = infiltration force, C_I = infiltration coefficient, F_S = static force, h = the depth of water column over the riverbed. Other parameters are defined in Figure 3.1.

It is appropriate at this time to introduce some qualitative information regarding streambed conditions observed in the field at various sites, with regards to saturation and bed shear resistance. Prior to initiation of pumping at the RBF facility at LWC, the riverbed immediately adjacent to the wellfield was inspected by a diver and found to be soft and easily penetrated. After one year of pumping, the same location was inspected and was observed to be highly resistive to penetration. While retrieving a sample, the diver had difficulty penetrating the riverbed surface with his knife, but when a sediment sample was removed by prying it from the riverbed, the sample became non-cohesive and free-flowing. The diver also observed that the cloudy water created by the sampling quickly disappeared into the hole where he had extracted the sample. The layer of shear-resistant riverbed was observed to be approximately 3 cm thick.

Similar conditions have been observed at the Greater Cincinnati Water Works Bolton RBF facility under high-pumping conditions, where the riverbed could not be sampled because of the shear resistance of the surface (Gollnitz 2004). The recharge basins in Sonoma County (California) Water Agency have exhibited similar behavior when under unsaturated conditions, with the flooded sand basins being highly resistive to penetration, but samples were totally non-cohesive when extracted. Similar observations have been made in high-rate filters in surface water treatment plants, when filter headloss exceeds the head over the filter (personal observation).

With this as a background, the force diagram was re-considered with the possibility of unsaturated conditions existing in the aquifer under a clogged riverbed. Under unsaturated conditions, static forces must be added to the overall forces acting on the particle or river surface. This additional force, and the balancing resistive force from the riverbed, is shown in Figure 3.2.

The clogging layer in Figure 3.2 is simplistically shown as a single membrane overlying the riverbed media. It is more likely that the conditions required to develop unsaturated conditions under a riverbed develop over time (months or years), and include particle straining at the surface, particle invasion into the riverbed media, and biological and geochemical processes. Thus, the saturated clogging layer above the unsaturated portion of the aquifer might be better represented as a zone upon which surrogate hydraulic forces are exerted, as opposed to forces on individual particles. The thickness of this saturated zone of riverbed overlying the unsaturated zone of the aquifer has been observed in the field by in-situ piezometers to be less than 0.6 m at Louisville, and in column studies to be less than 5 cm. Observations from divers at this site indicate that this zone may be as thin as 3 cm.

Comparative Analysis of Riverbed Forces

To put these forces into perspective, the typical range of each of these forces needs to be calculated. The science of sediment transport was well developed in the 1970's, and typical shear stresses on riverbeds have been reported in the range of 1 to 100 Newtons/square meter (N/m^2). Values of 20 N/m^2 are reasonable, as a horizontal force imparted by the moving stream over the riverbed surface.

In a given stream, the viscosity and particle density factors will remain constant regardless of the direction of flow, so that the forces imparted by an infiltration flow of water would differ from the horizontal forces primarily by fluid velocity, although an argument could be made that anisotropic conditions resulting from the sorting of sediment on the riverbed would impact the particle cross sectional area facing the flow vector. Approach velocities (Darcy velocity) in RBF systems have been measured with maximum values in the range of 0.09 m/hr or 0.00003 m/s (Wang, Hubbs, and Song 2002), compared to typical horizontal stream velocities in the range of 1.5 m/s and typical velocities at the streambed of 0.15 m/s. It is thus reasonable to conclude that the forces exerted by the vertical velocity component into the riverbed will be over three orders of magnitude lower than the horizontal shear force. Assuming a porosity of 0.2, the interstitial velocities would be roughly 5 times the approach velocity, or 0.00015 m/s. The forces exerted by the approach velocities and interstitial velocities are thus negligible compared to the horizontal shear forces.

The head loss exerted across a partially clogged riverbed can be directly calculated based on the average head loss across a given thickness of the streambed. Assuming a gradient of 0.5 across a 1 m thickness of media, the pressure exerted 1 m below the riverbed is 0.5 m of water, or approximately 5,000 N/m^2 .

Finally considering the total static force exerted by a column of water on the riverbed over an unsaturated aquifer, the force on the riverbed is the weight of the water column, and the resisting force must be provided by the riverbed itself. The pressure for a 5 m column of water is approximately 50,000 N/m^2, or roughly 2500 times greater than the shear stress exerted by the flowing stream on the riverbed. Approximate values for these forces are summarized below for comparison:

- Vertical Infiltration: 0.002 N/m^2
- Riverbed shear stress: 20 N/m^2
- Clogged infiltration (gradient = 0.5) 5,000 N/m^2
- Hydrostatic pressure: 50,000 N/m^2

The pressure gradient across a streambed over an unsaturated aquifer has been measured in the field at Louisville to be as great as 5 m headloss across a 0.6 m flowpath (the shortest distance measured), or greater than 8. Research in column studies designed to better define this gradient at Louisville indicated 5 m headloss across 0.05 m, or a gradient of approximately 100. This indicates that under unsaturated conditions, the forces exerted on the riverbed are comparatively huge, and can have a dramatic impact on the shear resistance and conductivity of the clogged riverbed.

STAGE AND FLOW HYDROGRAPHS

One of the most significant parameters in assessing the hydrology of a stream is bankfull discharge. Bankfull discharge is important because it can be used to approximate the dominant discharge (Knighton 1998), which is the flow responsible for transport of most of the sediment over time. If the bankfull discharge is responsible for most of the sediment transport, it is reasonable to infer that it would likely be most responsible for renewal of clogged areas in RBF systems.

The runoff regime of a river affects yield in RBF systems by altering the riverbed through erosion, transport, and deposition. The runoff regime can be characterized by parameters such as average discharge, average maximum yearly discharge, and average minimum yearly discharge near the RBF site. The maximum yearly discharge is computed by averaging the maximum yearly flows for all the years of record. Likewise, the minimum yearly discharge is computed by averaging the minimum yearly flows for all years of record. While these parameters enable rough comparison of flows from different streams, actual flow measurements in a time series are needed to understand the flow dynamics in a stream. A flow hydrograph (plot of discharge versus time) for a typical year can be used to illustrate the magnitude, frequency, and timing of runoff events.

A stage hydrograph (plot of water elevation versus time) is sometimes used in lieu of a flow hydrograph to characterize runoff dynamics. In natural streams, the relationship between stage and flow is fairly consistent and can be characterized by a stage/discharge function. The stage/discharge relationship often is less defined in regulated streams.

The stage of the stream determines the direction of flow between the stream and the aquifer. When the elevation of the surface of the stream is less than the elevation of the potentiometric surface in the aquifer, groundwater will flow from the aquifer into the stream (referred to as a gaining stream). When the surface of the stream is above the elevation of the potentiometric surface in the aquifer, the stream will recharge the aquifer (referred to as a losing stream). In RBF systems, drawdown from pumping lowers the phreatic surface in the aquifer to a level below the stream surface resulting in recharge from the stream to the aquifer. This type of stream-aquifer condition is referred to as a "leaky confined" aquifer.

Stage and Flow Hydrographs-Louisville

The flow in the Ohio River is managed through a series of dams along its entire length, and except during periods of high flow the pool at Louisville is maintained at an elevation 128 m above mean sea level (amsl). Bank-full flows of approximately 6 m above normal pool occur at a frequency of about once every four years. The 100 year flood is estimated at 138 m amsl, and the flood of record occurred in 1937, cresting at 140 m amsl, 12 m above normal pool. The daily

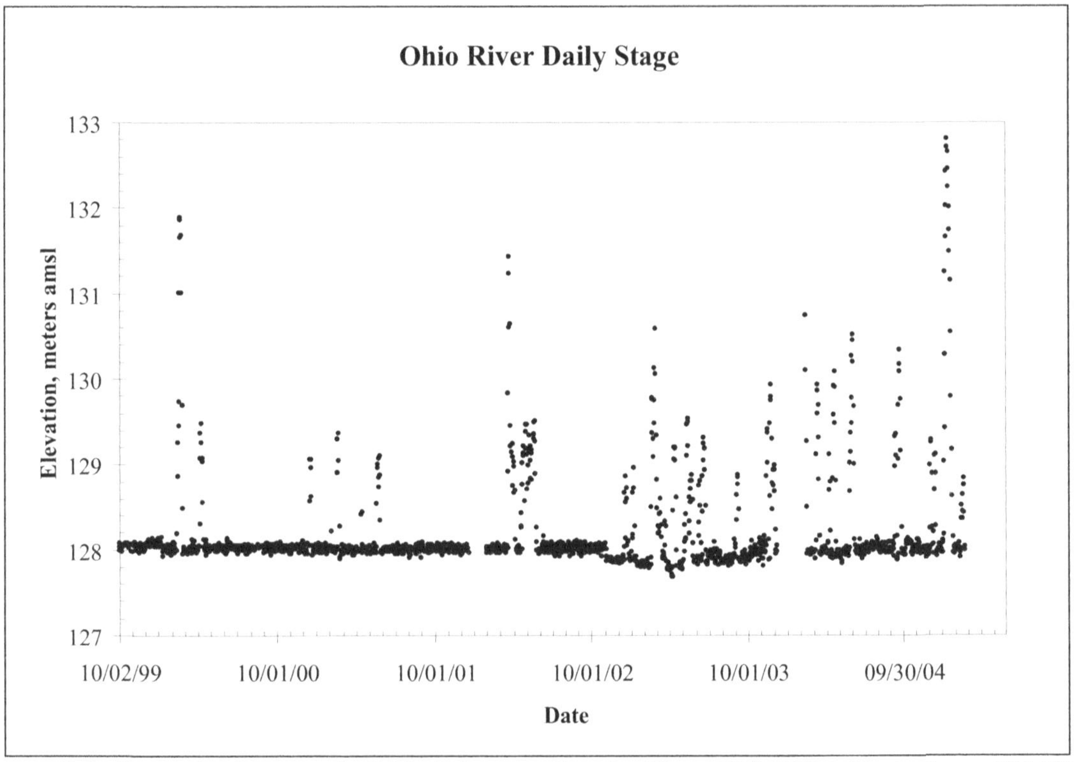

Figure 3.3 Stage hydrograph-Ohio River at Louisville: 1999-2004 (Data from USGS)

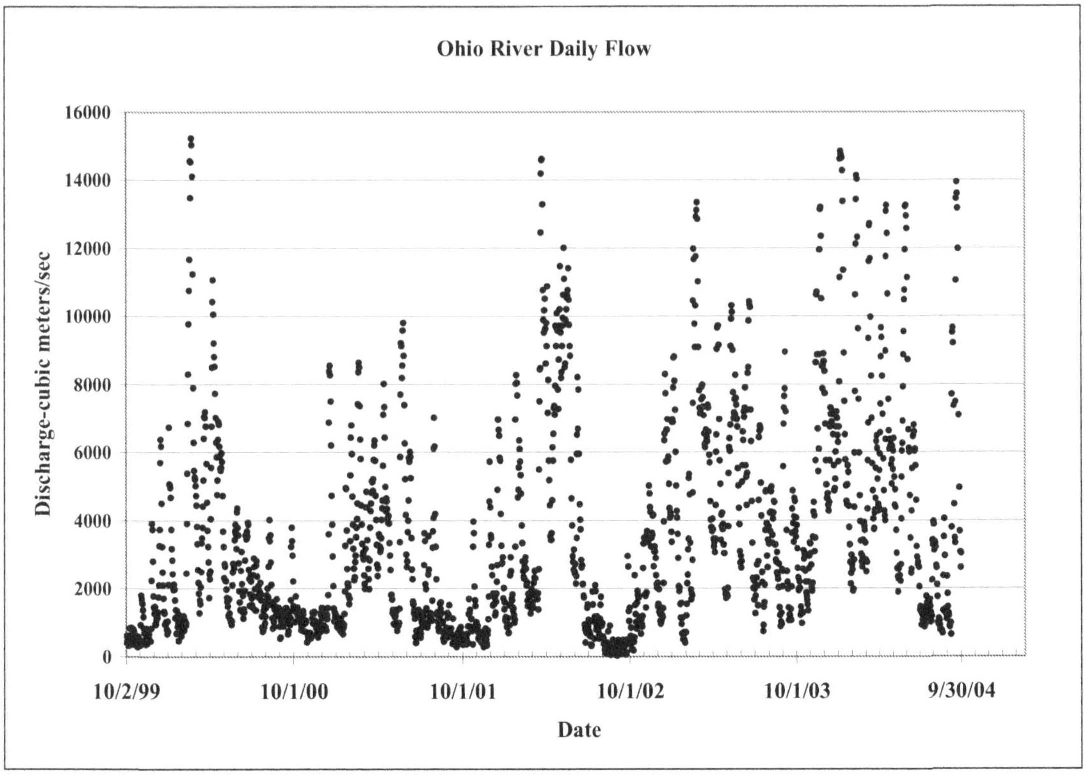

Figure 3.4 Flow hydrograph-Ohio River at Louisville, 1991-2001 (Data from USGS)

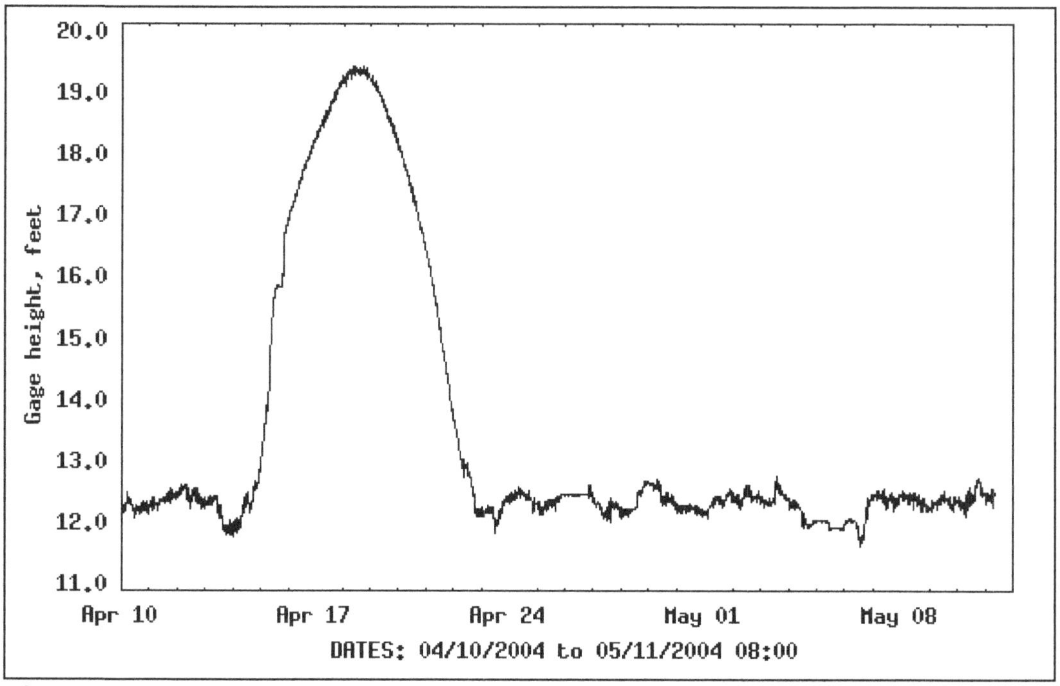

Figure 3.5 Flood surge stage hydrograph, Ohio River at Louisville (Data from USGS)

stage hydrograph is shown in Figure 3.3 for the water year 1999 through 2004.

The average daily flows for the Ohio River at Louisville are provided in Figure 3.4 for the period 1999-2004. Figure 3.3 and Figure 3.4 show that the Ohio River is characterized as a large, stage controlled river with occasional stage increases above pool. The flow hydrograph compared to the stage hydrograph illustrates the independence between flow and stage, except during conditions of high flow. Because of the artificial impact of the dams on river stage, the flow hydrograph better depicts the characteristic of the river with relation to RBF systems, particularly when considering sediment transport and riverbed scour.

A flood surge stage hydrograph for the Ohio River is presented in Figure 3.5. This particular hydrograph indicates a rapid rise in water level, indicating a steepened stream slope and associated increase in streambed shear stress and sediment transport. This type of stream event is particularly effective in moving streambed sediment, and would thus be expected to result in an increase in streambed scouring and renewal of the riverbed surface. Increases in specific capacity have been observed in Louisville following such events (Hubbs 2006). "Flashy" streams, or ones that experience rapid increases and decreases in stage, are typical of streams with high rates of sediment transport and riverbed scour.

RIVERBED SHEAR STRESS AND STREAM SLOPE DATA

The shear stress exerted on a riverbed can be estimated by several parameters, including the stream slope, stream velocity profile, and by an analysis of the streambed sediment in transport. Techniques for using stream velocity profiles and streambed particle analysis are presented in Appendix B. The simplest technique for estimating shear stress involves stream surface slope, which is presented in this section.

The total energy in a stream is the sum of the velocity head, pressure head, and elevation head. If, in a reach, most of the head loss is caused by friction along the riverbed and other energy losses are negligible (loss of momentum in bends, energy loss around obstructions, wind effects, etc.), the riverbed shear stress can be estimated through the change in energy between two adjacent cross sections. Assuming that the change in velocity head is small and the stream has a uniform cross section, the head loss is approximately equal to the difference in elevation of the water surface between the two sections. Thus, the water surface slope in a given reach is an indicator of the shear stress imposed on the riverbed. Recognizing the limitations of the assumption of uniform flow, the average shear stress can be estimated from the water surface slope as follows:

$$\tau_0 = \gamma * R * s \quad \textbf{(1)}$$

In equation (1), τ_0 is the mean boundary shear stress, R is the hydraulic radius (closely estimated by stream depth in larger streams), s is the water surface slope, and γ is the specific weight of the fluid (Knighton, 1998). This equation will tend to overestimate the shear stress in natural channels since the energy loss is not limited to friction losses alone. For simplicity, the water surface slope has been utilized as a surrogate for riverbed shear stress for the rivers in this study.

River profiles were generated for each selected site by plotting the surface elevation of the river at average flow versus distance along the river. These profiles show the slope of the river and the location of the RBF system along the river. From the profiles shown in Figure 3.6 and Figure 3.7, it can be seen that the RBF sites consistently are located along the middle to lower reaches of the rivers. One of the most important distinctions among the rivers is the steepness of the profiles. The river morphologies range from long rivers with relatively flat slopes to short rivers with comparatively steep slopes. The profiles are presented in two categories for visual clarity. Rivers longer than 1000 km or higher than 500 m in elevation are shown in Figure 3.6. The Platte River, a relatively short river, is included in this graph to show its confluence with the Missouri River. The remaining rivers are displayed in Figure 3.7. A constant horizontal to vertical aspect ratio was used so that the slopes of the rivers shown on the two graphs can be directly compared visually.

SITE GEOMETRY

The site geometry of RBF facilities impacts the ability of the river to recharge the aquifer, and influences the severity of clogging. The location of a wellfield can be described by its distance from the river and proximity to river bends. The perpendicular distance from the center of the wellfield to the point of recharge in the river defines the distance across which the head loss in the aquifer is distributed. Assuming that riverbed head losses and wellscreen head losses are similar, the RBF design with a greater setback from the riverbank will yield less water, as the distance of travel is greater.

RBF systems can be located on straight river sections or along the inside (convex side) or outside (concave side) of a river bend. The riverbed on the outside of a bend experiences greater velocities and momentum changes and thus greater shear stress for a given discharge than does the riverbed on the inside of a bend. Ironically, bed areas on the outsides of bends can be more susceptible to clogging. According to Schubert (2006), large pieces of rock or large pieces of

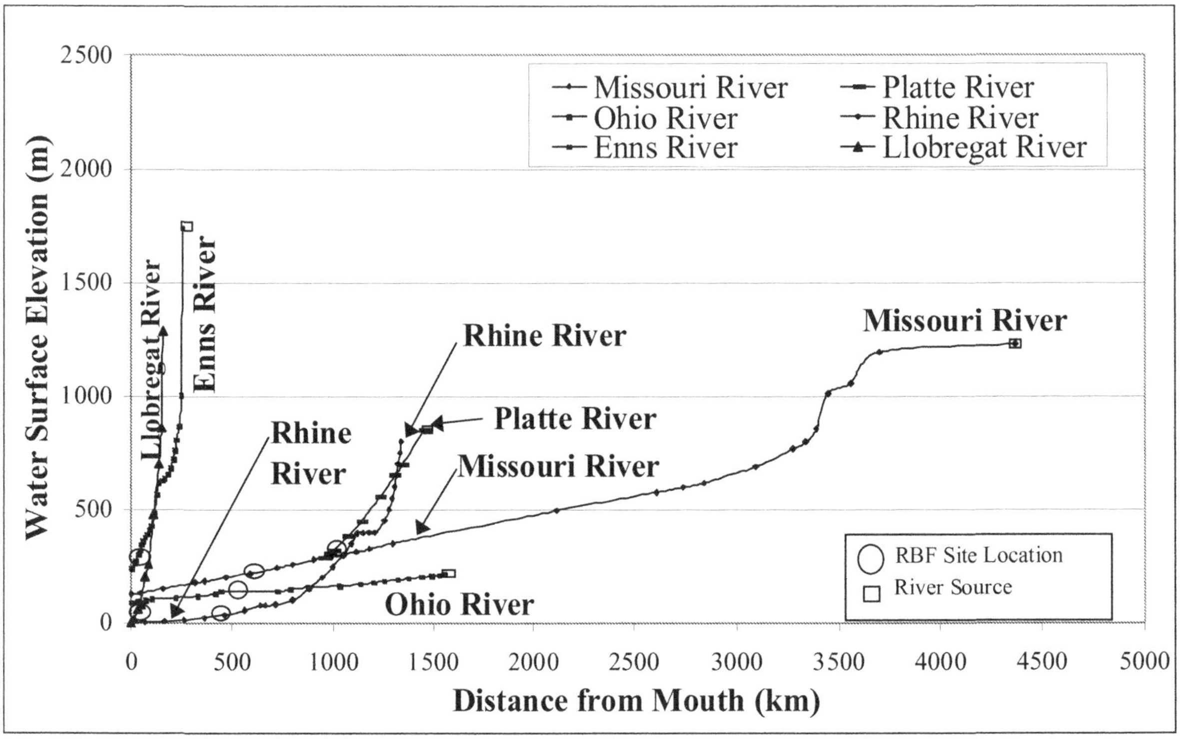

Figure 3.6 River profiles for longer rivers

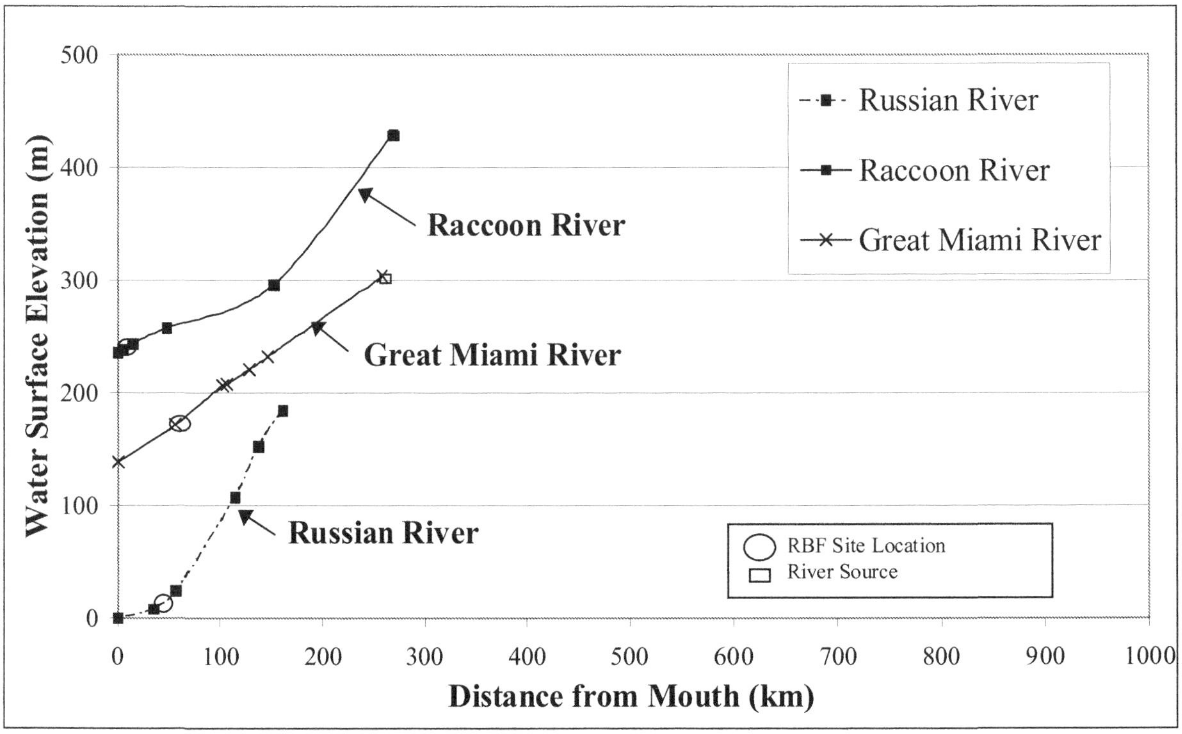

Figure 3.7 River profiles for shorter rivers

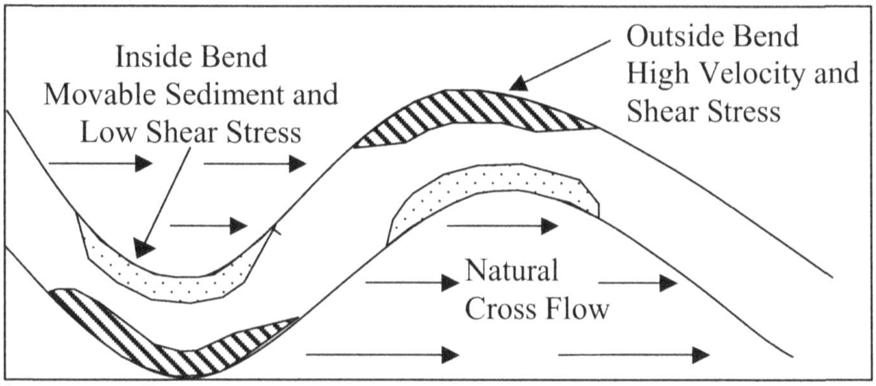

Figure 3.8 River meander

debris deposited by high velocities along the outsides of bends may form armor layers and shield underlying clogged layers from the renewal effects of shear stress during passage of flood waves. Lower river velocities along the insides of bends can remove materials in clogged zones not protected by an armor layer of large stones. Thus, the clogging that occurs along the inside of a bend may be more likely to be disrupted by flood waves. Another advantage in locating wellfields at the insides of bends is that the natural groundwater flow across such areas results in additional recharge to the aquifer (Schubert 2006). See Figure 3.8.

The layout of the wellfield affects yield by controlling the flow paths of the river water through the riverbed to the well. Wellfields with multiple wells spaced along the river impact a longer length of riverbank than do single wells, and thus experience recharge from a longer reach in a river. However, the length of river impacted per well tends to be smaller in wellfields with multiple wells unless the wells are spaced so far apart that the zones of influence of the wells do not overlap. The total length of river impacted by all the wells in the system is expected to significantly influence yield. The methodology used to estimate the length of river impacted for rivers in this study is described in Chapter 2.

The geometry (width and depth) of a river can also influence yield in RBF systems. Specifically, in a river where clogging is problematic, a large width can act to compensate for capacity by allowing the infiltration area to spread across a wider area. In narrow rivers, the infiltration area can extend across the entire width of the river. In such a situation, clogging has a more significant impact on yield, since the width of the infiltration area cannot increase to compensate for lost capacity. Conversely, when clogging occurs in RBF systems along a wide river, the width of the infiltration area grows to partially compensate for clogging. This phenomenon is described in Chapter 4 for a pump test of a radial collector well in Louisville. Since the depth and, to some extent the width, of a river change with stage, the depth and width of a river at average discharge and at bankfull discharge has been used to characterize the geometry of rivers at RBF sites in this study.

ESTIMATING SHEAR VELOCITIES CAUSED BY RIVER TRAFFIC

The focus of this chapter has been the estimation of shear stresses caused by natural streamflow. However, on streams used for commercial transport, the bed shear stresses exerted by a passing vessel can exceed even the highest stresses exerted by natural flow. In regularly traveled streams, this traffic can represent a short-term, high frequency shear stress event that could have a significant impact on mediating the impacts of clogging.

The techniques for measuring the magnitude of shear stress from a passing vessel include direct measurement of velocity near a passing vessel and measurements from scale-models of rivers using simulated vessels. While data from these two techniques are sometimes conflicting, it is obvious that vessels on smaller navigable streams exert significant shear stresses on stream bottoms. (Maynard, 1990 and 2000)

Several elements of vessel passage causing shear stresses have been described in the literature. The first is a reversed-pulse resulting from the headwater immediately in front of the vessel, which leads the vessel by as much as 50 m. Depending on whether the vessel is traveling upstream or down, the drag force from the vessel can impart a higher or lower velocity on the streambed. The displacement of water under the vessel can cause significant local changes in stream velocity (up to 1 m per second), especially when the submerged vessel cross-section is large compared to the stream cross section. Behind the passing vessel, the prop-thrust impact tends to be exerted primarily on the stream surface, with return flow occurring at the streambed.

At the point immediately below the propeller, however, the return flow re-circulating from behind the tow and the upstream flow being accelerated through the propeller meet, creating a significant riverbed stress. At this location, flow directions have been observed to be vertically upward towards the propeller. It is anticipated that this upward flow would result in a highly turbulent flow pattern, and that significant shear stress would be anticipated on the riverbed.

It has also been observed that riverbed shear forces are much greater when vessels are changing directions or starting motion. Shear stresses under these conditions can be much more significant than for a passing vessel at constant velocity.

The published information is augmented with the author's observation of visually noticeable increases in stream turbidity behind a passing vessel on the Ohio River, as observed from the air. During low turbidity river conditions, the area of higher turbidity was sharply defined as a narrow plume approximately the width of the vessel, and was visible for at least one km beyond the passing vessel. It is thus concluded that passing vessels can exert adequate shear forces in their immediate path capable of re-suspending riverbed sediment, and could have an impact in the overall clogging of riverbeds.

FIELD DATA

The wellfields included in the study displayed a wide range of hydrologic, hydraulic, and physical site characteristics. The well-to-river distances were between 35 m and 260 m for the majority of the sites. The well to river distance for the Llobregat River was more than five times greater than for the other locations. Well spacing varied from 20 m for the Staad and Flehe wellfields to 1000 m in the Platte River Wellfield. The Grind wellfield on the Rhine River had the largest total length of wellscreen. The hydrologic and hydraulic data relevant to yield in RBF systems are summarized in Table 3.1. The Site Geometry information was split into two tables because of size limitations (Table 3.2 and Table 3.3).

Table 3.1 River hydrology/hydraulics factors

River	Wellfield	River Discharge: Avg. (m^3/s)	River Discharge: Bank-full (m^3/s)	River Discharge: Avg. Max. Yearly (m^3/s)	River Discharge: Avg. Min. Yearly (m^3/s)	Bankfull: Cross-sectional Area (m^2)	Bankfull: Mean Vel. (m/s)	Average Discharge: Cross-sectional Area (m^2)	Average Discharge: Mean Vel. (m/s)	Average Discharge: River Surface Slope (m/km)
Llobregat	Cornellà	20		400	2			5*	1.5*	0.8
Rhine	Grind	2100		9400	810			2000	1.0	0.2
Rhine	Flehe, PW V	2100		9400	810			2000	1.0	0.2
Rhine	Staad, PW I	2100		9400	810			2000	1.0	0.2
Enns	KRB02	160		900	70			300	0.6	1.8
Ohio	BE-PAYNE	3300	8500	14000	240	6700	1.3	6000	0.5	0.04**
Raccoon	Maffitt	50	300	500	4	200	1.7	60	1.0	0.5
Missouri	Nearman	1500	5100	4800	410	3000	1.8	1000	1.2	0.2
Great Miami	Bolton	100	900	1000	10	500	1.8	200	0.4	0.7
Russian	Wohler / Mirabel	40†	500‡	800†	5†		2.0§		0.6§	0.8
Platte	Ashland	160	3700	1000	30	1000	2.6	200	0.8	0.2

* Data shown is for most frequent flow. The most frequent flow is 8 m^3/s

† Discharge measurements were computed by adding the daily flow data from USGS gage 11465350 at the mouth of Dry Creek to daily flows from USGS gage 11464000 near Healdsburg approximately 19 km upstream of wellfield. Dry Creek flows were not measured for discharge values greater than 5.7 m^3/s. Missing flow records were filled in using flow values of 5.7 m^3/s. This method should result in an underestimation of flows from Dry Creek, and will affect the average flow, average maximum yearly flow, and the average minimum yearly flow.

‡ Bankfull discharge given represents the bankfull condition of Russian River near Healdsburg 19 km upstream of the wellfield.

§ Velocity at average and bankfull discharge was estimated using USGS measurement data from gage 1464000 near Healdsburg because discharge data for the wellfield location were unavailble. The condition likely differs from that of the wellfield.

** Wellifeld is located in pool upstream of a dam. Data given represent the pool slope.

Table 3.2 Site geometry characteristics – 1

River	Wellfield	Distance from River to Wells (m)	Location Relative to River Bends	Length of River Impacted (m)	Number of Wells	Distance Between Wells (m)
Llobregat	Cornellà	1650	Unaffected	Negligible	26 Extracting Wells, 7 recharging wells	200
Rhine	Grind	260	Inside Bend	5000	7 Radial Collectors	500
Rhine	Flehe, PW V	50	Outside Bend	1400	70 Vertical Filter Wells Connected by Siphon Pipe	20
Rhine	Staad, PW I	35	Straight	500	25 Vertical Filter Wells Connected by Siphon Pipe	20
Enns	KRB02	48	Inside Bend	500	Single Vertical Well	N/A
Ohio	BEPAYNE	37	Straight	600	Single Radial Collector	N/A
Raccoon	Maffitt		Straight		5 Radial Collectors and 1 Horizontal Well	
Missouri	Nearman	61	Straight		Single Radial Collector Well	N/A
Great Miami	Bolton	100 *	Varies	2500	10 Vertical Wells	250
Russian River	Wohler Bridge and Mirabel	10-50	Straight/Inside † Bend	3000	5 -Radial Collectors	300
Platte	Ashland	130	Island in River	1500	2 - Radial Collectors	1000

* Distance provide for low flow condition

† The Wohler Bridge wellfield is on a straight section of the river. The Mirabel wellfield is on an inside bend.

Table 3.3 Site geometry characteristics - 2

River	Wellfield	Average Length of Wellscreen (m)	Total Length of Wellscreen (m)	Depth of River at Average Discharge (m)	Depth of River at Bankfull Discharge (m)	Width of River at Average Discharge (m)	Width of River at Bankfull Discharge (m)
Llobregat	Cornellà	10	260	0.5 *		25	25
Rhine	Grind	670	4700	12	20	380	
Rhine	Flehe, PW V	19	1300	12	20	380	
Rhine	Staad, PW I	19	480	12	20	380	430
Enns	KRB02	2	2	5		80	90
Ohio	BEPAYNE	480	480	10	16	610	640
Raccoon	Maffitt	210 †	1100 †	1	3	60	90
Missouri	Nearman			4	8.2	305	351
Great Miami	Bolton	17	170	2	4	60	150
Russian River	Wohler Bridge and Mirabel	128	5646	1-4	12	60	80
Platte	Ashland	400	800	1	5	300	400

* Depth given is for the most probable discharge.

† Wellscreen information displayed for the Maffitt Wellfield on the Raccoon River is for the five radial collector wells only. Data were not provided for the inclined horizontal well.

CHAPTER 4
RIVERBED CHARACTERISICS

The riverbed is the interface between the river and the aquifer. Thus, the riverbed influences the ability of the river to recharge the aquifer, as is measured by the hydraulic conductivity of the riverbed. The hydraulic conductivity of a riverbed can be estimated by analyzing the particle size of the riverbed media, or computed by measuring the head loss and flux rate across a given distance into the aquifer. Hydraulic conductivity can also be estimated using computer modeling, assuming that all other aquifer and well parameters are known, and solving for an average streambed hydraulic conductivity.

The riverbed hydraulic conductivity in a given RBF system would be expected to vary both in time and space. As sediment is deposited on and in the riverbed, the hydraulic conductivity would be expected to decrease as a function of the amount of sediment deposition resulting from water flow through the riverbed. As this flux rate varies with head, and thus distance from the well, the deterioration in riverbed conductivity would be expected to lessen with distance from the well. Seasonal impacts of high riverbed shear and riverbed surface renewal during flooding events result in step-increases in conductivity, the extent depending on the degree of clogging and the intensity of the scouring event.

In streams with high amounts of biodegradable organic carbon, bioclogging can reduce the conductivity deep into the bed and below the typical scouring depth, resulting in a more permanent loss of conductivity. This condition has been observed in streams impacted by substantial wastewater influence (Schubert 2006). A more detailed discussion of bioclogging in porous media is provided by Engesgaard (2006), Baveye et al. (1998), and Vandevivere and Baveye (1992).

In systems where the design and operation results in partial dewatering of the aquifer under the riverbed, the development of unsaturated conditions under the riverbed results in dramatic decreases in riverbed conductivity. This condition essentially shifts the point of recharge further from the well, increasing the head loss required for a given well discharge and decreasing overall system capacity. This condition has been observed at Louisville, and field data is provided in this chapter and Appendix D.

Riverbed hydraulic conductivity is thus a highly complex and variable parameter, which can be the limiting factor in the capacity of an RBF system. This parameter is the least understood and most difficult to estimate in the design of RBF systems. This chapter provides an overview of techniques for estimating riverbed conductivity, detailed data from the RBF site at Louisville, and field data from participating utilities.

FLUX RATES, HEAD LOSS, AND HYDRAULIC CONDUCTIVITY-BASICS

The calculation of hydraulic conductivity requires a known headloss across a defined flow path at a given discharge flow rate. These data are easily determined under laboratory conditions, but are difficult to obtain in the field. Thus, estimates of these values are used to calculate riverbed conductivity in the field. The discharge rate, or Darcy velocity, can be estimated using tracers such as temperature (Constantz, 2006) or directly measured using

Figure 4.1 Core sample of riverbed sediment taken from the Missouri River at confluence with Mississippi River. Core diameter is 3.75 cm (1.5 inches).

seepage meters. The flow path distance is often estimated as the vertical distance between the riverbed and the point of head measurement. However, flow paths in RBF systems are often complicated, with both vertical and horizontal components. Thus, a flow path estimated assuming vertical flow would represent the minimum flow path distance estimate, and would thus represent the minimum velocity for the calculation of hydraulic conductivity. The head loss across a given flow path may be measured directly (using piezometers), or estimated from measured or modeled points in the system.

Average values for streambed conductivity can be estimated by the grain distribution of the riverbed (Domenico and Schwartz 1998; Freeze and Cherry, 1979). The variation of grain size distribution in the vertical direction, however, can result in significant reduction in hydraulic conductivity. This is illustrated in Figure 4.1 for a riverbed core sample taken from the Missouri River near St. Louis (approximately to scale). A layer of fine sediment overlays fine sand, on a deeper bed of course sand.

Computer models can be used to estimate riverbed conductivity, assuming that aquifer values for conductivity and well characteristics are known. Calibration of these models requires data on water level and temperature both land-side and under the riverbed, as well as well discharge. These data from under the riverbed are often unavailable, thus limiting the quality of the estimate from the model. This estimation technique usually yields an average value that provides the best fit to available data. Assigning a lower conductivity to this parameter near the well has provided better fit to available data (Wett 2006).

RIVERBED CHARACTERISTICS AT LOUISVILLE

Many Midwestern U.S. rivers have sand bottoms and mud banks of fine silt and clay. The extension of the mud layer into the river has an obvious impact on riverbed conductivity, essentially extending the point of recharge into the river. This mud layer can vary significantly, both spatially and temporally. At Louisville, the mud layer extends at least 20 m from the river edge, and varies by as much as 1 m in depth depending on recent streamflow events. The RBF wellfield plan and profile at Louisville is provided in Figure 4.2, showing the approximate location of the mud layer and monitoring points used in this research. Each lateral is approximately 70 m in length. Lateral L4 is approximately 15 m below the riverbed at its end. Piezometers P39 and P37 were set approximately 0.6 and 2.7 m below the riverbed directly over the end of lateral L4. The depth to bedrock from land surface at the caisson is approximately 30 m, and the sand aquifer is approximately 25 m thick.

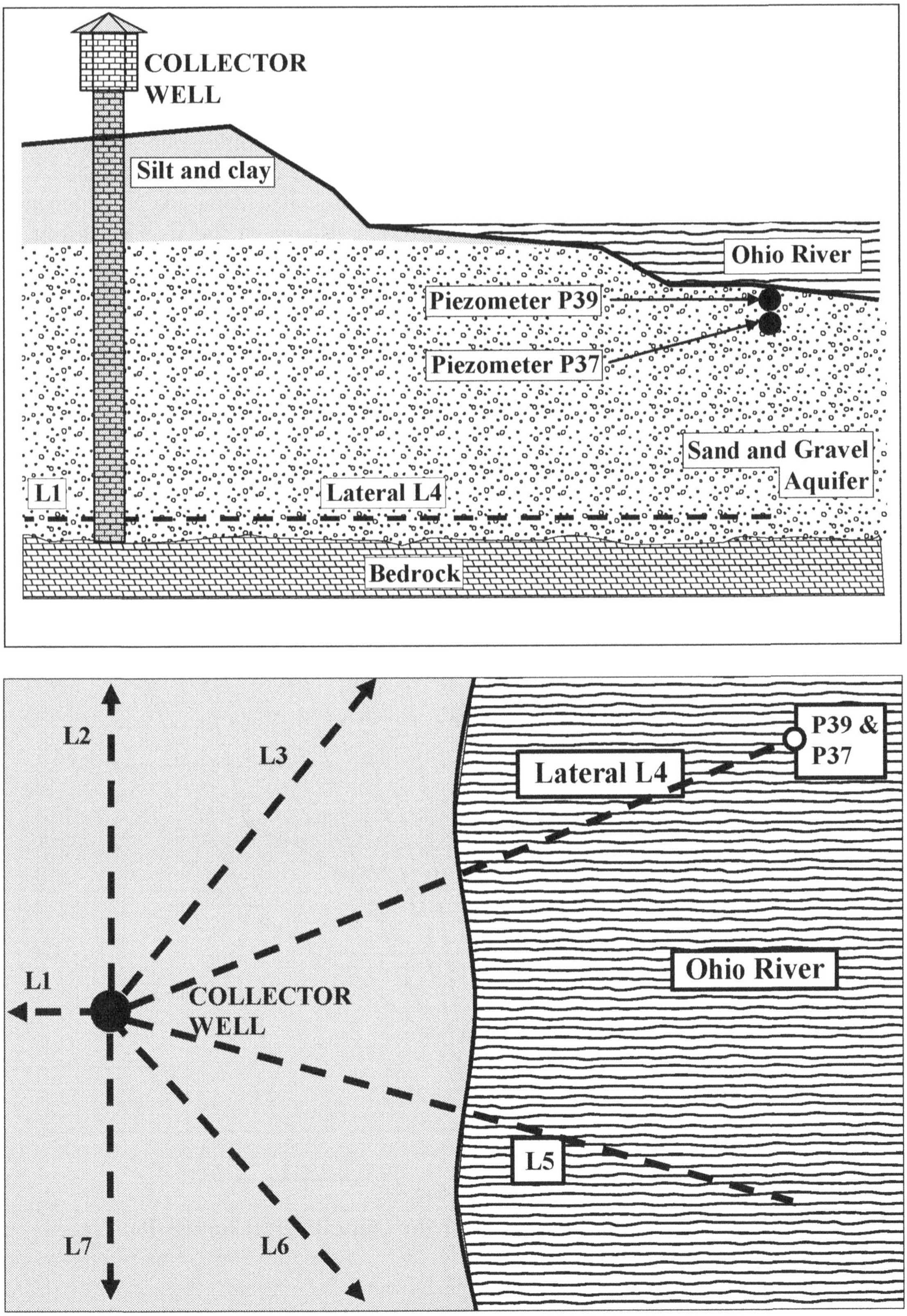

Figure 4.2 Plan and profile of LWC RBF facility (profile through lateral L4).

Typical particle size distributions for suspended sediment and the riverbed are presented in Figure 4.3. Data for the sediment load in suspension and in the riverbed are provided for typical flow conditions. These data characterize the Ohio River as a highly turbid stream with a fairly consistent riverbed media of medium to fine sand. Very little cobble is found on the riverbed at this site. Those areas of the stream characterized by lower streambed scouring velocities (backwaters and shallow areas) are typically associated with smaller sediment sizes. Figure 4.3 also shows the particle size distribution of a sample of the riverbed taken from a clogged area adjacent to the well in the summer of 2002. This particular sample was extremely shear resistant, and piezometer data indicate that the aquifer beneath the riverbed was likely unsaturated at the time of sampling. This sample was collected by a diver, who noted that the sample was extremely shear resistant (not yielding to a knife). When extracted, however, the sample became non-cohesive. Approximately 65% of this sediment sample was between 0.1 and 0.25 mm, and approximately 10% of the sample was less than 0.1 mm.

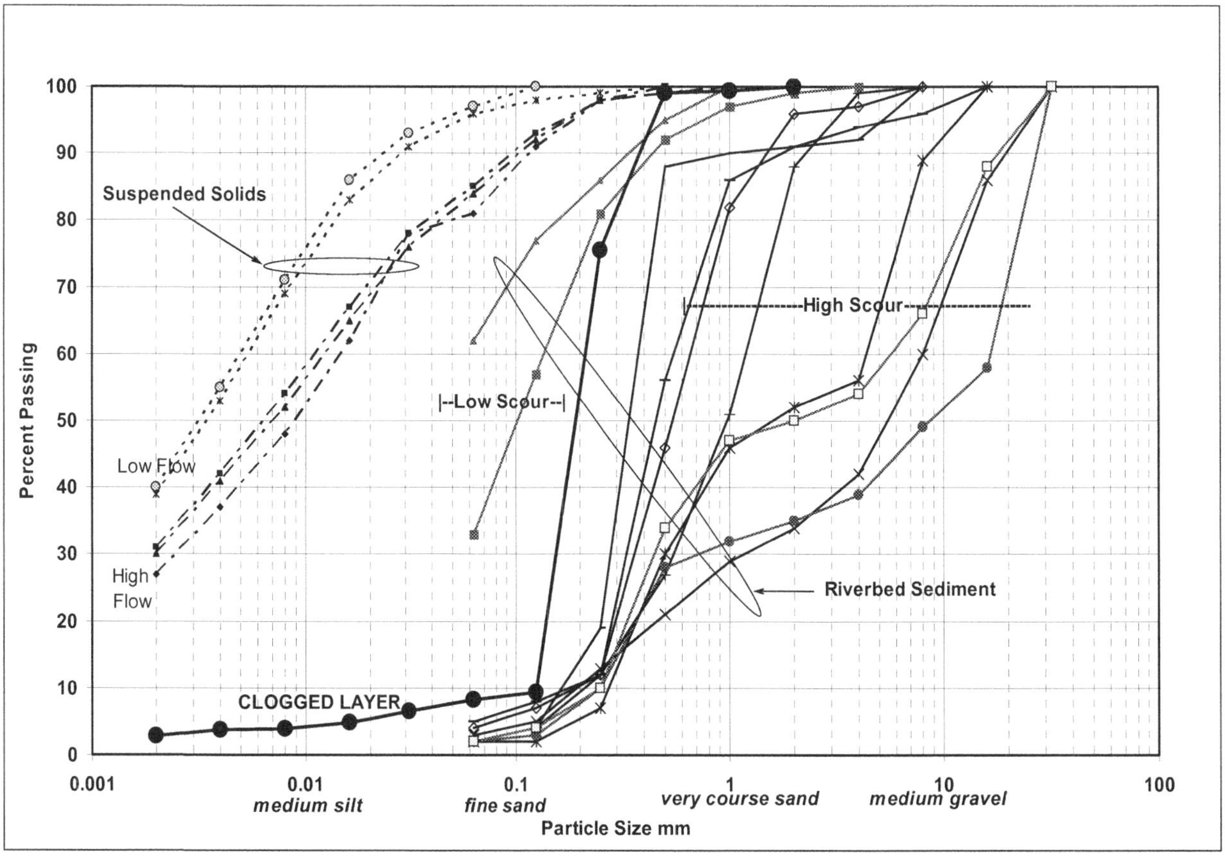

Figure 4.3 Stream sediment characteristics for the Ohio River at Louisville.
Data taken for riverbed sediment at rivermile 602 from the 1980 USGS annual report. Suspended solids data were taken at rivermile 602 from USGS annual reports from 1979-1982. Data from the clogged layer were collected by LWC and analyzed by USGS (August 2002).

Table 4.1 Comparison of D_{90}, D_{50}, and D_{10} particle sizes for the riverbed and aquifer at similar elevations, LWC RBF facility on the Ohio River.

	Aquifer	Riverbed
D_{90}	12.35 mm	11.7mm
D_{50}	0.72 mm	2.1mm
D_{10}	0.28mm	0.2mm

A comparison of the D_{90}, D_{50}, and D_{10} particle sizes for the riverbed and the adjacent aquifer is provided in Table 4.1. The aquifer sample was taken from a core sample approximately 30 m from the bank, at the same elevation as the riverbed. These data indicate that the riverbed and aquifer are similar in composition.

HEAD LOSS ACROSS THE RIVERBED AT LOUISVILLE

Since the initiation of pumping in the summer of 1999 and through 2004 data were collected from piezometers placed in the riverbed immediately over the end of lateral collector L4 (Figure 4.2). The original installation included three Geokon model 4500 vibrating wire piezometers placed approximately 0.6 m (P39), 1.5 m (P38), and 2.7 m (P37) below the riverbed. These unvented probes were capable of reading positive and negative pressure and temperature in the aquifer. The probe 1.5 m below the riverbed failed in the winter of 2001, but the remaining two piezometers continued to operate through the duration of this study.

In the summer of 2002 additional pressure and temperature transducers were set in the riverbed perpendicular to the bank at distances approximately 120, 200, and 275 m from the well (shown as data points on the piezometric curves in Figure 4.4). The probes were set approximately 1.5 m below the riverbed, and were functional for only a short period of time, becoming disabled with the flood event in the spring of 2003. These probes allowed the shape of the piezometric surface under the river to be measured under various operational conditions.

Data for summer 2002 and winter 2003 from these probes are presented in Figure 4.4. Of particular interest is the location of the piezometric surface in conjunction with the shape of the riverbed. During warm weather conditions, the piezometric surface intersects the riverbed surface approximately 67 m from the well at a well discharge of 68,000 m^3/d (18 mgd). During cold weather conditions and a well discharge of 64,000 m^3/d (17 mgd), however, these two surfaces intersect as far as 90 m from the well. If the piezometric surface were to drop slightly lower than shown in Figure 4.4, the intersection of these two surfaces could extend out as far as 120 m from the well. The impact of this lowering of the piezometric surface on the dewatering of the aquifer and resulting decrease in hydraulic conductivity will be further discussed in Chapter 4.

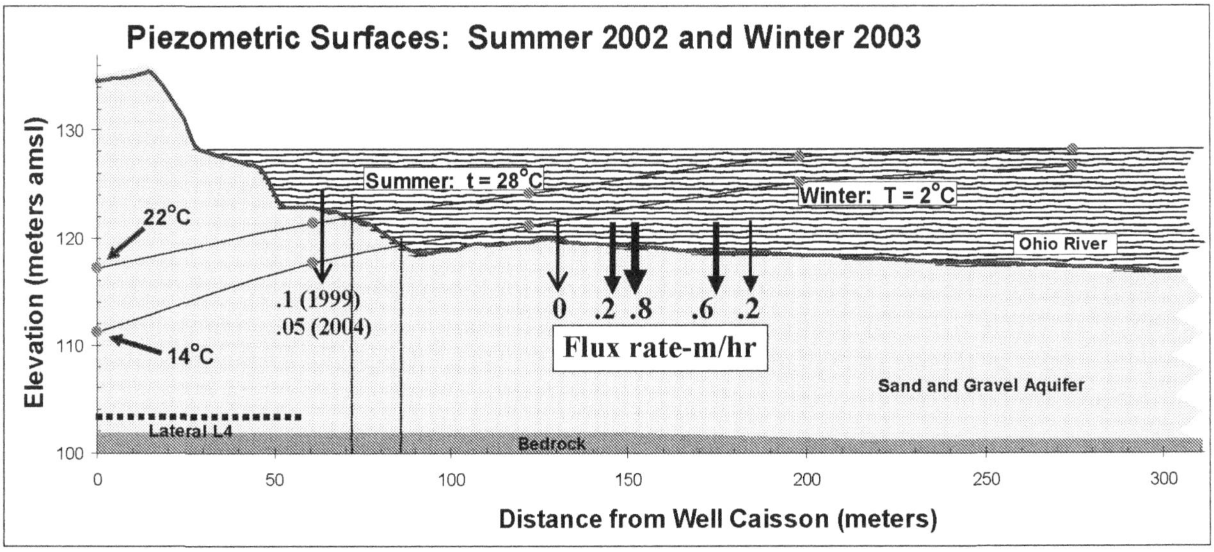

Figure 4.4 Piezometric surface and flux rates.
Measures of piezometric surface were taken summer 2002 and winter 2003. Flux rates taken Summer 2004.

FLUX RATES ACROSS THE RIVERBED AT LOUISVILLE

Flux rates across a riverbed can be directly measured by seepage meters designed for the specific riverbed application, or inferred from temperature data at the start-up of RBF systems. Both techniques were used at Louisville, with reasonable agreement between the two techniques.

Flux Measurements Estimated by Temperature Profiles

Data from the initial start-up of the well in 1999 and the repeat of this test in July 2004 were used to estimate riverbed flux rates by analyzing the temperature breakthrough curve between the warmer river water and the colder aquifer over lateral L4 at temperature probe depths of 0.6 and 2.7 m below the riverbed (Figure 4.2). This technique provided an estimate of the vertical velocity component in the aquifer, which was then adjusted by porosity to provide an estimate of the Darcy velocity. This technique can provide an accurate in-aquifer measure of vertical velocity, but can only be used when there is a significant temperature difference between the river water and the aquifer water at the point of measurement. These conditions typically exist at the beginning of system pumping, or after a several month period of no pumping. Calculated flux values for these temperature profiles are presented in Figure 4.4 (directly over the end of lateral L4) and later in this section with the calculations for riverbed hydraulic conductivity.

This technique also requires that the aquifer be fully saturated and the water velocity be primarily vertical in order to determine the length of the flowpath for determining velocity. If the flow velocity has a significant horizontal component, the result will be an underestimation of the flow path distance, and thus an underestimation of the fluid velocity. Thus, assuming that all flow is in the vertical direction will result in the computation of a minimum value for riverbed hydraulic conductivity; any deviation from this assumption would result in the calculation of a higher value of hydraulic conductivity.

Flux Rates Measured by Seepage Meter

Direct measurements of the flux rate in the riverbed at Louisville were taken with a seepage meter designed for use in deep rivers. The prototype device consisted of a 55 gallon drum cut in half and affixed to a hose. The device was placed by a diver and left on the riverbed for several days before being lost to river traffic. Based on this experience, a heavier, smaller device was designed that could be easily placed on the riverbottom from a workboat. The device included a weighted round steel hood with an area of 0.093 m^2, with an attached hose allowing a bladder to be connected at the river's surface. This device worked adequately in depths up to 10 m (see Figure 4.5).

Figure 4.5 Seepage meter

The accuracy of the measurement was significantly impacted by river velocities, wind action, and waves. Under ideal conditions (river velocity under 0.3 m/hr, low wind, and no waves), reproducible data were obtained. As conditions deteriorated, so did the quality of the data obtained. A camera attached to the device allowed observation of the integrity of the seal between the device and the riverbed during the monitoring process.

A limiting condition for the riverbed flux measurement was the erosion of the seal between the monitoring device and the riverbed. Under higher river velocities, the sand around the upstream face of the monitoring device would gradually erode away, exposing the inside of the hood directly to the stream. This resulted in a gradual decrease in measured flux rate with time, to a point of no measured flux. Thus the data were scrutinized, and data that were believed to have been influenced by the erosion of the seal were not included in the averages presented below. Data from several other monitoring locations were either very low, or no seal could be obtained between the riverbed and the device, and were eliminated from further analysis. All data are provided in Appendix C.

Initial data taken in the summer of 2003 with a prototype device indicated that discharge velocities as high as 0.3 m/hr might be expected in zones of high infiltration. These values were considerably higher than those measured in 1999 using the temperature profile technique. Assuming an aquifer porosity of 0.2, the maximum discharge velocity measured at the location of the probes in 1999 was 0.1 m/hr. It is noted that the direct flux measurements were taken at a location where infiltration was believed to be high based on information on the riverbed hardpan from a diver and piezometric data. This location was approximately 150 m from the well; the temperature profile data were taken over lateral L4 approximately 60 m from the well.

Selected flux data from the seepage meter taken during August, 2004 are listed in Table 4.2 and graphically presented in Figure 4.4. The monitoring locations are presented as distance from the well perpendicular to the riverbank, with the exception of the location at 127 m, which

Table 4.2: Direct riverbed flux measurements, August 2004.

Radial distance from well, meters	Discharge velocity, meters/hour
127 (1)	0.0
146	0.23
150	0.82
175	0.55
183	0.23

(1)-This point is 75 m downstream of lateral L4.

was 75 m downstream of the well. Pumping rates were held constant at about 64,000 m^3/d (17 mgd) throughout the monitoring period. River conditions were acceptable on some of the days of monitoring and marginal on others. The data presented represent those data felt to be the most reliable, based on multiple measures at the same point, and are believed to best represent the riverbed flux rates for the given monitoring point.

A deliberate attempt was made to locate the point of highest flux in the riverbed, and the data indicate that under the conditions existing in August 2004, this point was between 146 and 175 m from the well, perpendicular to the riverbed at the location of the well. The maximum measured flux rate was 0.82 m/hr at a point 150 m from the well. This location was beyond the area where the hardpan was observed by the diver and beyond the point where unsaturated conditions under the riverbed were indicated.

In conjunction with the flux measurements taken above in late August 2004, an attempt was made to characterize the hardness of the riverbed. In August 2002 a diver had probed the riverbed with a knife and identified the hardpan as extending as far as 150 m from the well. This hardpan was assumed to be the response of the riverbed to unsaturated conditions below. This hardpan was also indicated by "blow counts" from the setting of temporary piezometers in the river.

MEASURES OF RIVERBED HYDRAULIC CONDUCTIVITY AT LOUISVILLE

Flux rates derived from direct measures and inferred from temperature breakthrough curves were used to calculate the hydraulic conductivity of the riverbed according to Darcy's equation as follows:

$$K_{rb} = \text{(discharge velocity)/(headloss gradient)}$$

For temperature-estimated flux measurements, discharge velocity was calculated by dividing the flowpath distance (between the riverbed and the probe) by the travel time (measured at the mid-point of the temperature breakthrough curve), and multiplying this by the porosity of the formation (assumed to be 0.2). The headloss gradient was calculated as the difference between total head at the riverbed (river surface elevation) and the total head measured at the probes divided by the vertical flowpath distance.

Hydraulic Conductivity from Temperature Data

Given the above assumptions and limitations, data from June 1999 and July 2004 were used to estimate riverbed hydraulic conductivity over the end of Lateral L4. Prior to the 2004 test, the pressure of each probe and the difference in pressure between the two probes were compared to readings for similar non-pumping conditions and river level in 1999, and were found to be remarkably consistent over the 5 year period of in-situ use (within 0.03 m of water head over the 5-year period).

The well discharge flow rate during the June 1999 study was 76,000 m^3/d (20 mgd) compared to 64,000 m^3/d (17 mgd) in 2004. River and aquifer temperatures were also different in the two periods: in 1999, river temperature was 26.5°C, while the aquifer temperature at 3 m below the riverbed was 15.1°C. The temperatures at these same sample locations were 28.6°C and 9.9°C, respectively before the start of the 2004 test. The colder aquifer temperature in 2004 was likely the result of the aquifer being charged with cold river water prior to the cessation of pumping in April 2004.

A significant scouring event occurred shortly after the well was taken out of service in April 2004, and it is likely that the riverbed over the probes was deeply scoured and fresh sediment was deposited over the area between April and July. This erosion/deposition pattern is consistent with bathemetry data from 2004 and observations from previous years.

Data used to estimate hydraulic conductivity for 1999 and 2004 are presented Table 4.3. It is noted that in 1999, the temperature at 0.6 meter and 2.7 m below the surface reached the same temperature as the river water (26.5°C) within 7 hours. For the 2004 data, these temperatures reached equilibrium only after 13 days.

Riverbed hydraulic conductivity values were estimated from these data. It was necessary to assume that for the very early period of initial pumping the velocity was predominantly vertical. Otherwise, the distance required to compute velocity would be unknown. This assumption was likely more valid in 1999 than in 2004.

Table 4.3: Field data and calculations for riverbed hydraulic conductivity.

Year	Distance (m)	Headloss (m)	Time (hours)	Headloss (m/m)	Velocity (m/h) (5)	Riverbed K (m/hr)	Riverbed K (cm/sec)
1999	0.6	1.87	1.30	3.07	0.094	0.031	.0009
2004 (1)	0.6	3.64	**2.5**	5.97	0.05	**0.008**	**.0002**
2004 (2)	0.6	4.17	5.2	6.84	0.02	0.003	.0001
2004 (3)	0.6	5.53	70	9.08	0.002	0.0002	.000005
1999	2.7	2.15	5.7	0.79	0.09	0.12	.003
2004 (4)	2.7	4.72	**10.4**	1.75	0.05	**0.03**	**.0008**

(1)-headloss and velocity from the first 2 hours of the pump test; (2)-headloss and velocity from hours 2 through 8 of the pump test; (3)-headloss and velocity from hours 12 through 18 of the pump test; (4)-headloss and velocity from the first 10 hours of the pump test; (5)-discharge velocity was calculated assuming aquifer porosity of 0.2. Riverbed aquifer temperature at start of test 1999 was approximately 5°C warmer than in 2004. River temperature was approximately 2°C colder in 1999 than in 2004.

The data from 1999 were easily interpreted, as the shapes of the temperature breakthrough profiles at 0.6 m and 2.7 m below the riverbed were as would be expected for non-varying conditions of hydraulic conductivity. The 1999 data indicated a mean vertical velocity of 0.46 and 0.53 m/hr at the 0.6 and 2.7 m depths, respectively, indicating good agreement between the two datasets.

The data from 2004, however, were distinctly different from those of 1999, with indications that the velocity through the riverbed was neither constant nor vertical during the 13 days required for temperature levels to stabilize. The data from the first 20 hours of the 2004 test were used to estimate vertical velocity, however, as these data indicated periods during which the temperature breakthrough curves were consistent and somewhat interpretable, with the flow velocity vector assumedly in a primarily vertical direction. Beyond 20 hours, the temperature breakthrough displayed trends indicative of significant changes in riverbed conductivity. The data and calculations are provided in Table 4.3 for riverbed hydraulic conductivity between the riverbed and the first 0.6 m and 2.7 m below the riverbed, for data from 1999 and 2004.

The velocity values for the earlier part of the 2004 dataset are approximately half that observed in 1999. However, at approximately 11 hours after the start of pumping, temperature data indicate dramatic changes in velocity at 2.7 m below the riverbed for the 2004 dataset. Given only gradual changes in head, it is inferred that these dramatic velocity changes were the result of a dramatic change in riverbed hydraulic conductivity in the vertical flow path.

The earliest portions of the datasets for the 2004 data are most comparable to the 1999 data for comparison of changes in hydraulic conductivity, in that they show the least indications of transient changes in riverbed conductivity. These are the data from the first 2.5 hours for the 0.6 m depth (K_{rb}=0.002 m/hr), and the first 10.4 hours for the 2.7 m depth (K_{rb} =0.008 m/hr) for the 2004 test. For calculations from the 0.6 m and 2.7 m flowpath distances, the riverbank conductivity in 2004 was approximately 25% of the value from 1999. The reader is referred to Appendix B for a more rigorous analysis and interpretation of these data.

Using data for head loss between the probes P37 and P39, and the velocity estimates in Table 4.3, vertical aquifer conductivity was calculated between 0.6 and 2.7 feet in the aquifer, exclusive of the head loss in the first 0.6 m of the riverbed. These values were 1 m/hr for the 1999 dataset and 1.5 m/hr for the 2004 dataset. These data compare favorably with the horizontal conductivity value of 5 m/hr for the entire aquifer independently derived by Schafer (2000) using traditional pumping tests in at the site. Horizontal to vertical conductivity ratios in the range of 3 to 1 might be expected in alluvial aquifers.

Comments: Measures of Riverbed Hydraulic Conductivity with Temperature at Louisville

The accuracy of measures of riverbed hydraulic conductivity using temperature as an indicator are subject to assumptions made regarding flow direction, accurate depth measurement between the probes and riverbed, and accurate measures of pressure and temperature. The assumption of vertical flow is acknowledged as being violated shortly after the start of pumping as the cone of depression is developed and the piezometric surface dips towards the well. As flow varies from vertical to having a horizontal component, the affect would be to underestimate the length of the flow path between the riverbed and the point of measure, thus underestimating the velocity. Noting the relative flatness of the piezometric surface under typical operating conditions (measured at a slope of approximately 1/20 horizontal/vertical at the location of the probes), the magnitude of the vertical head loss (up to 6 m loss per m of vertical travel), and the similarity between the steepest slopes of the temperature breakthrough curves in 1999 and 2004, the early data from these pump tests provide reasonable estimates of riverbed hydraulic conductivity.

The distance between the probes can be inferred by the pressure measured from them during non-pumping conditions. This distance has been measured within 2 cm, or within 1% of the approximate 2.14 m distance between them. Calculations under non-pumping conditions using probe pressure also identify the depth of the probe below the river surface with similar accuracy.

The distance between the riverbed and the probes, however, is both poorly measured and highly variable. The probes were initially set at distances assumed to be 0.6 m and 2.7 m below the riverbed. The riverbed near the location where the probes were set has been measured to vary in depth by more than 2 m from spring to fall. This variation could have had a significant impact on the calculation of riverbed hydraulic conductivity in both 1999 and 2004.

Data from 2004 indicate that a significant change occurred when the probes were reading a water head of approximately 123 m amsl. This is approximately the elevation of the riverbed at that point and time. The riverbed elevation has been observed to vary between 120.8 and 123.3 m amsl near the location of the probes, with the higher elevations typical of later summer/early fall. This assumption is consistent with the bathemetry indicating that the riverbed at this location aggrades from spring to fall. This observation is significant with later discussions regarding unsaturated conditions under the riverbed.

Riverbed Conductivity Measured from Seepage Meter

For direct flux measurements, the Darcy (discharge) velocity was measured directly from the seepage meter at various points in the aquifer. The maximum value for riverbed hydraulic conductivity was calculated at a point approximately 130 m from the riverbank and perpendicular to the well. To compute a hydraulic conductivity from the flux values, an estimate of hydraulic gradient was needed. This estimate was made by interpolating between the points of the summer 2002 piezometric surface (Figure 4.4) as measured 1.5 m below the riverbed, adjusted for the pressure reading from piezometer P37 for the day of flux measurement. While this estimate is crude, it was considered adequate for an order-of-magnitude estimate of hydraulic conductivity. This yielded a head loss estimate of 2.5 m across a distance of 2.7 m for a gradient of 0.9m/m, and a flux rate of 0.8 m/hr. This was the highest flux value measured during the 2004 monitoring event. The calculated riverbed hydraulic conductivity under these conditions and assumptions was 0.9 m/hr. This value is significantly higher than that measured by the temperature profile technique for start-up conditions in 1999 measured 30 m from the bank at 0.12 m/hr.

IMPACTS OF LOSS OF AQUIFER SATURATION UNDER THE RIVERBED ON RIVERBED CONDUCTIVITY AT LOUISVILLE

The operational data collected from start of pumping included water levels in and surrounding the well and under the riverbed, temperature in the river and well, and well discharge. In addition, several "snapshot" datasets in which well discharge and river water levels were constant for extended periods of time provided operational data used for calibrating various models. These data, along with the data from the 1999 and 2004 pump start-ups, provided a unique opportunity to evaluate changes in riverbed hydraulic conductivity after 5 years of operation, and to illustrate the impact of unsaturated conditions developing in the aquifer under the riverbed on the hydraulic conductivity of the riverbed. These data are presented and discussed in the following sections.

Evaluating unsaturated conditions under the riverbed

A review of data from probes P37 and P39 indicated that the pressure in probe P39 (0.6 m below the riverbed) very closely followed the pressure in P37 when adjusted for the 2.1 m difference in elevation between the two probes, except when the pressure in P37 dropped below 2 m of water. The vibrating wire piezometers used in this study were capable of measuring negative pressures, and thus if the water column between P39 and P37 remained saturated, a negative pressure would be expected whenever the pressure in P37 dropped below 2.1 m of water. Conversely, unsaturated conditions would be indicated if the pressure in P39 was near atmospheric pressure whenever P37 dropped below 2.1 m of water.

Data from July 2004 through December 2004 were analyzed to examine the difference in pressures between P37 and P39 (Figure 4.6). The x axis in this figure represents the pressure at the lower probe (P37), and the y axis represents the difference in pressure between the two probes. The difference in pressure between these probes would be expected to be nearly constant (approximately 2.1 m of water) when the aquifer was saturated, and would be expected to be a function of the piezometric surface at P37 if the aquifer in the area of the upper probe (P39) became unsaturated. The change in correlation of the pressure differential between P39 and P37 at the point where the potentiometric surface dropped below the elevation of P39 (pressure at P37 less than 2.1 m of water) indicated that the aquifer at probe P39 had become unsaturated.

Pumping conditions were controlled in May 2005 to further investigate the transition that occurs when the pressure at probe P37 approaches 2.1 m of water and the pressure at P39 approaches atmospheric pressure. Pumping was adjusted initially to establish a slowly declining pressure at P37 between 2.2 and 2.0 m of water, followed by an increase in pumping adequate to drop the pressure in P37 to approximately 1 m of water. Pressures in both probes were closely monitored during a period of 4 days. These data are presented in a time series in Figure 4.7.

It is noted that probe P39 shows a negative pressure as the water level drops below 2.1 m in probe P37. Over a span of two days, however, this negative pressure tends to increase towards atmospheric pressure despite the continued decrease in pressure at P37. At mid-day May 15 the pumping rate was increased an additional 7000 m^3/d (2 mgd) to drive the potentiometric surface even lower. The initial response was that the pressure at both P37 and P39 dropped; however, after approximately 4 hours the pressure drop in P39 stopped, and reversed with a trend towards atmospheric pressure despite the continued drop in pressure at P37. These data are consistent with conditions expected if the aquifer at probe P39 were draining and transitioning to an unsaturated state.

These data are considered supportive of the hypothesis that under conditions when the total head measured in the aquifer drops below 122 m amsl (approximately 2 m of water pressure head at P37), the aquifer at probe P39 undergoes transition from saturated to unsaturated, and the total head at P37 represents the piezometric surface in the aquifer under the riverbed. The aquifer does exhibit a temporary transition to negative pressure as the potentiometric surface drops below the pressure probe, but over the period of several days, the aquifer drains towards

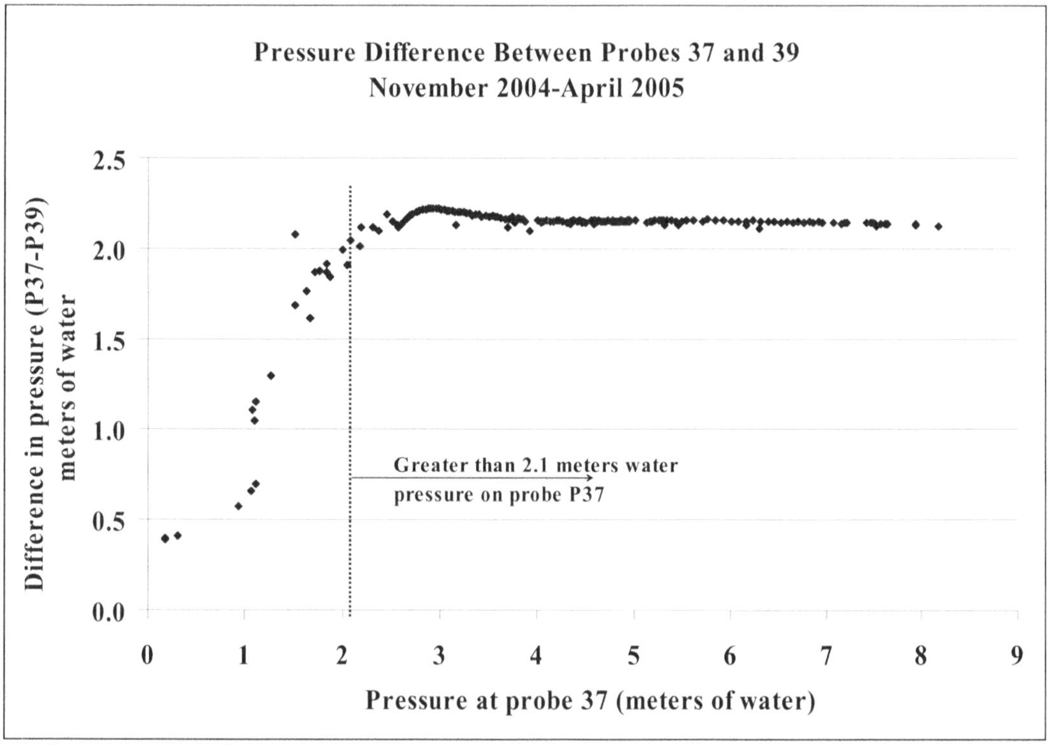

Figure 4.6 Pressure difference between piezometers P37 and P39 as a function of pressure at piezometer P37.

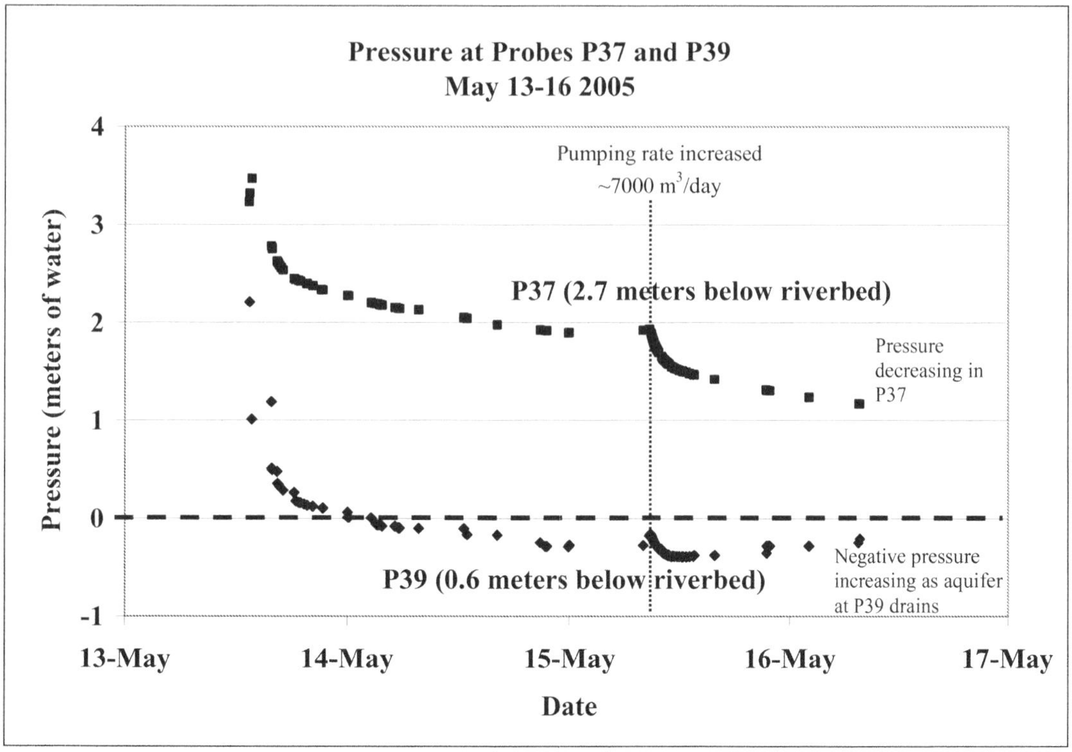

Figure 4.7 Transition between saturated and unsaturated conditions.

the well and pressures balance, resulting in a near-atmospheric pressure reading at probe P39, 0.6 m below the riverbed. The 0.5 m negative pressure observed at probe P39 is consistent with the anticipated capillary pressure in sand as it drains (Domenico and Schwartz 1998).

A detailed evaluation of data from pump tests conducted in 1999 and 2004 with regards to the impact of unsaturated conditions developing under the riverbed is provided in Appendix D. Data from the Louisville site were considered in developing the following hypothesis for riverbed clogging when unsaturated conditions develop in an RBF system.

Hypothesis for riverbed clogging under unsaturated conditions

The reduction in riverbed conductivity over time can be attributed to particle impingement on the surface, particle invasion into the riverbed matrix, biological clogging, and geochemical clogging. Another clogging mechanism is offered: that the top few centimeters of riverbed composed of a mixture of clay, silt, and sand loosely deposited in alluvial processes exhibit relatively high permeability until the aquifer supporting them becomes unsaturated, at which time the forces of the overlying water column compress the loose soil into a structure with much lower hydraulic conductivity. This, in conjunction with the decreased conductivity related to loss of saturation, results in a dramatic decrease in hydraulic conductivity.

If unsaturated conditions exist under the riverbed, the entire weight of the water column is supported by the riverbed above the unsaturated zone. It is reasonable to expect that as unsaturated conditions develop, the portion of the aquifer above the unsaturated zone would drain, forcing the headloss of the overlying water column across a thinning upper saturated zone. As this de-saturation progresses, the weight of the overlying column of water is transferred from the entire submerged aquifer to a relatively thin layer of the riverbed. As this proceeds, the conductivity of the riverbed decreases dramatically.

While the thickness of riverbed assuming this load is unknown, it is known to be less than 0.6 m at Louisville (the location of the pressure probe closest to the riverbed). When a diver successfully penetrated this layer with his knife, he observed it to be less than 2 cm thick, and upon penetration substantial flow into the riverbed was observed as the muddied water disappeared into small hole created in the riverbed. Thus, it appears that at this point of observation very near the edge of the unsaturated zone, the thickness of the layer of very low hydraulic conductivity was in the range of 2 cm thick. Preliminary column work ongoing at Louisville further indicates that 97.7% of all headloss across a 0.6 m deep column occurs in the top 5 cm of filter media under saturated conditions. Following a period of partial column desaturation similar to what is observed in the field, the headloss in the top 5 cm of filter increased to 99.7%. The visibly observed layer of fine sediment was approximately 2 cm thick on the top of the column.

This hypothesis is consistent with observations in RBF facilities in Louisville, Sonoma County, and Cincinnati Ohio.

Given the potential for a dramatic decrease in riverbed conductance as hypothesized above, the importance of riverbed area, width, and aquifer storage become critical elements in a wellfield design. Systems with wide streams, massive storage and highly transmissive formations may be able to withstand periods of high demand by drawing from aquifer storage without having the piezometric surface dip below the riverbed. However, systems with smaller streams and aquifers closely confined by valley walls and shallow depth to bedrock have a higher reliance on infiltration, and may find that under conditions when capacity is needed the most, riverbed conductance could decrease dramatically as a result unsaturated conditions developing under the riverbed, reducing overall system capacity. This condition is exacerbated under conditions of low water temperature, as the increased viscosity of water drives the piezometric surface even lower for a given flow.

FIELD DATA FROM PARTICIPATING SITES

Riverbed media data from participating sites are presented in Table 4.4. The wide range of values for riverbed conductivity was attributed in part to differences in methods of defining and measuring riverbed conductivity. With the exception of the Enns River, the D_{10} media values were relatively consistent, varying between 0.2 and 0.4 mm. Data for D_{50}, and D_{90} values were more variable, ranging from 0.5-20 mm and 1.9 to 30 mm, respectively. An armor layer was reported at five of the study sites.

Table 4.4: Riverbed characteristics

River	Wellfield	Riverbed Hydrauilic Conductivity (m/day)	D_{10} (mm)	D_{50} (mm)	D_{90} (mm)	Armor Layer ?
Llobregat	Cornellà	? *				Yes
Rhine	Grind	260 †		18		No
Rhine	Flehe, PW V	300 †	0.4	12	30	Yes
Rhine	Staad, PW I	300 †		20		Yes
Enns	KRB02	2 ‡	0.022	2		No
Ohio	BEPAYNE	15 §	0.25	2	8	No
Raccoon	Maffitt					
Missouri	Nearman		0.3	0.6	1.6	
Great Miami	Bolton	0.5 **				Yes ††
Russian	Wohler Bridge and Mirabel		0.3	4.4	17	Yes ‡‡
Platte	Ashland	80 §§	0.2	0.5	1.9	

* Under normal conditions a layer of silt completely blocks the infiltration of water and the riverbed is fully clogged. In the case of flooding, the silt layer is removed and for a short period of time the riverbed conductivity is similar to the aquifer conductivity.

† Riverbed conductivity values for the Rhine River were derived from particle size distributions for samples extracted from several borehole cores.

‡ Value refers to conductivity of riverbank. The riverbed conductivity is 0 m/day. The value for the riverbank conductivity was estimated by calibrating a numerical model and assuming a 1-m thickness of riverbank.

§ Estimate from "Evaluation of Riverbank Filtration as a Drinking Water Treatment Process" for the top 0.6 m of riverbed medium (Wang, Hubbs, and Song, 2002, 66)

** Conductivity was the highest value obtained from a pump test during the the design of the wellfield (Gollnitz, Whitteberry, and Vogt, 2004, 103)

†† An armor layer was observed during a site visit. However, there was evidence that the armor layer was frequently transported during flood events.

‡‡ There was evidence that the armor layer is moved frequently during flood events.

§§ Value provided by Lee (2005)

CHAPTER 5
AQUIFER CHARACTERISTICS AND WATER QUALITY

AQUIFER CHARACTERISTICS

Hydraulic Conductivity, Volume and Storage Basics

The aquifer properties that influence yield in RBF systems are the volume, transmissivity, storativity, and porosity of the aquifer. The volume of the aquifer is important because it determines the quantity of water available for abstraction from a well when recharge is minimal.

The volume of the aquifer is a function of the saturated thickness, width, and length of the aquifer. The length of the aquifer may not be determinate in alluvial aquifers where the aquifer follows the river path. Therefore, the saturated thickness and width of the aquifer alone will be used to characterize the quantity of water available per unit length of the river.

The transmissivity of an aquifer characterizes the ease with which water flows through the aquifer to the well. The transmissivity of the aquifer is a function of the conductivity of aquifer medium and the saturated thickness of the aquifer. Aquifer transmissivity values commonly are obtained from pump tests at RBF sites. The transmissivity is the rate of flow per unit width of aquifer through the entire thickness of the aquifer due to a unit hydraulic gradient. The product of estimated aquifer conductivity and saturated thickness can be used in lieu of transmissivity when transmissivity data from pump tests are not available.

The storativity of an aquifer is the amount of water that is released per unit area for a unit drawdown in the phreatic surface in an unconfined aquifer. The storage coefficient is the amount of water released from a confined aquifer per unit area for a unit reduction in potentiometric level. Thus, storativity partially characterizes the drawdown effects of the pumping well, particularly during periods of little or no recharge from the adjacent river. The porosity is the ratio of the volume of voids to the total volume of an element of aquifer. Porosity influences the velocity at which water moves through an aquifer under a given hydraulic gradient. However, a very fine-grained medium may have a relatively high porosity but a very low hydraulic conductivity because of poor connectivity of the voids and/or the attraction for water molecules of clay mineral particles.

Data from Participating Utilities

Data obtained for aquifer characteristics are presented in Table 4.5. Large, highly transmissive aquifers are expected to have positive influences on production in RBF systems. The aquifer at the Llobregat River had the highest transmissivity of all the sites. However, as mentioned previously, the RBF system on the Llobregat River is unique and difficult to compare with other systems. The systems with the larger aquifers included wellfields at the Great Miami and Rhine Rivers. The Enns River aquifer was the smallest of all the aquifers.

Table 5.1: Aquifer characteristics

River	Wellfield	Aquifer Width (River Side) (km)	Aquifer Width Total (WA) (km)	Sat. Thick-ness (Th) (m)	Aquifer Area (m^2)	Trans-missivity (m^2/day)	Stora-tivity	Poro-sity
Llobregat	Cornellà		2.5 *	40	100,000	18000		0.38
Rhine	Grind	5	10.0	15	150,000	3900		0.2
Rhine	Flehe, PW V	5	10.0	15	150,000	3900		0.2
Rhine	Staad, PW I	5	10.0	15	150,000	3900		0.2
Enns	KRB02	0.4	0.8	5	4,000	1600		0.2
Ohio	BEPAYNE	2	2.0	20	40,000	2500	0.0005	0.38
Raccoon	Maffitt	N/A †	N/A †	N/A †	N/A †	N/A †	N/A †	N/A †
Missouri	Nearman					9600		
Great Miami	Bolton	0.3	2.8	50	141,000	5700		0.25
Russian	Wohler / Mirabel			15		10600		
Platte	Ashland	1	2.6	20	53,000	3100	0.2	0.3

* Aquifer width shown for wellfield at Cornellà on the Llobregat River is given for location of RBF system. The aquifer width near the area of recharge from the river varies from 0.25 km to 2.1 km.

† No aquifer information was furnished for the Raccoon River site.

WATER QUALITY

Impact of Water Quality on System Capacity

River water quality affects well yield by influencing hydraulic conductivity of the riverbed. Decreases in the hydraulic conductivity of the riverbed can result from mechanical, biological, and/or chemical clogging. Therefore, the biological and chemical makeup of the water, and size of suspended sediment can influence yield in RBF systems.

As described previously, mechanical clogging occurs when suspended sediment in the river becomes trapped in the riverbed as water passes from the river to the aquifer. The potential for mechanical clogging can be described by the size and amount of particles in suspension (Schubert 2002a). The size of suspended sediment can be described by a particle distribution or through the use of key sizes such as D_{10}, D_{50}, or D_{90}. These sizes tend to increase as the velocities in the stream increases. Thus, suspended sediment size and concentration is influenced by discharge. The size distribution and concentration of suspended sediment at average discharge were chosen to represent the characteristics of the suspended sediment in the river. The amount of suspended sediment can be represented by the total suspended solids concentration, or estimated by the turbidity of the water.

Biological clogging occurs when a biological film forms in the riverbed due to the accumulation of microbial bodies and/or their metabolic by-products (Vandevivere and Baveye

1992, Baveye et al. 1998). In order for biological clogging to occur, there must be sufficient nutrients in the water to support biological activity (Engesgaard 2006). Total organic carbon (TOC), chlorophyll, and dissolved oxygen concentrations of the water are three factors that can be used to quantify the nutrient content and the metabolic support character of the water. Excessive amounts or poorly treated wastewater in the recharge stream can have significant impacts on riverbed clogging in RBF systems (Schubert, 2002a).

Chemical clogging is the loss of hydraulic conductivity of the riverbed due to clogging by chemical precipitants. Some factors thought to influence chemical clogging are the iron, ammonia, and nitrate concentrations, and the hardness of the water. Biological activity can also lead to chemical clogging when high levels of biodegradable material cause changes in the redox potential and pH of the river water which in turn results in the precipitation of substances (Schubert 2002a). Thus, in addition to the factors identified above, the pH of river water and extracted water should be considered when evaluating the potential for chemical clogging. Values at average discharge were chosen to represent the chemical quality of the selected rivers. Stuyfzand et al. (2006) provides additional information on chemical clogging of RBF systems.

Comparison of Water Quality Characteristics

Water quality parameters collected from participating RBF sites impacting biological, chemical, and mechanical clogging processes are provided in Table 5.2, Table 5.3 and Table 5.4. The water quality data were split into three tables because of space constraints.

Table 5.2: Water quality characteristics –1

River	Wellfield	Suspended Sediment At Avg. Discharge			Turbidity at Average Discharge (NTU)	Total Suspended Solids at Average Discharge (mg/l)
		D_{10} (mm)	D_{50} (mm)	D_{90} (mm)		
Llobregat	Cornellà				270	
Rhine	Grind	<.5 *	5 *	20 *	35	
Rhine	Flehe, PW V	<.5 *	5 *	20 *	35	
Rhine	Staad, PW I	<.5 *	5 *	20 *	35	
Enns	KRB02					16
Ohio	BEPAYNE	<0.002	0.005	0.03	60	40
Raccoon	Maffitt				50	180
Missouri	Nearman				130	500
Great Miami	Bolton				30	80
Russian	Wohler / Mirabel		0.08	0.09	22	
Platte	Ashland		0.3	0.12	50	330

* Data given is for estimated maximum size transferred or moved.

Table 5.3: Water quality characteristics – 2

River	Wellfield	Nutrients and Biological Activity at Average Discharge			
		Total Phos. (mg/l)	TOC (mg/l)	Dis. Oxygen (mg/l)	Chlorophyll Content (mg/l)
Llobregat	Cornellà		6	7.9	0.03
Rhine	Grind	0.22	3	10.6	0.009
Rhine	Flehe, PW V	0.22	3	10.6	0.009
Rhine	Staad, PW I	0.22	3	10.6	0.009
Enns	KRB02	0.05	2	10-12	
Ohio	BEPAYNE	0.2	3	10-12	0.005
Raccoon	Maffitt	0.3		9	
Missouri	Nearman		4	9.9	
Great Miami	Bolton	0.2	5	10	
Russian	Wohler / Mirabel	0.2	4	9	
Platte	Ashland	0.5	4	9.6	

Table 5.4: Water quality characteristics – 3

River	Wellfield	Chemical Clogging Parameters at Average Discharge				
		Nitrate (mg/l as N)	Amm - onia (mg/l as N)	Hardness (mg/l as $CaCO_3$)	Iron (mg/l as Fe)	pH
Llobregat	Cornellà	9.8	1	400	0.06	8.5
Rhine	Grind	3.6	0.2	250	1.2	7.9
Rhine	Flehe, PW V	3.6	0.2	250	1.2	7.9
Rhine	Staad, PW I	3.6	0.2	250	1.2	7.9
Enns	KRB02	4	0.05	170	0.5	8.2
Ohio	BEPAYNE	1.4	0.07	140	0.9	7.9
Raccoon	Maffitt	12	0.05	290	0.9	
Missouri	Nearman			230	4.5	8.0
Great Miami	Bolton				0.02	8.1
Russian	Wohler / Mirabel	0.2	0.05	100	2.5	7.6
Platte	Ashland	0.9	0.1	190	10	8.2

CHAPTER 6
EVALUATING WELL YIELD AND SPECIFIC CAPACITY

WELL YIELD, SPECIFIC CAPACITY, AND TEMPERATURE-EFFECTS BASICS

The amount of water available for extraction from a particular wellfield is influenced by all of the factors described in Chapters 3 through 5. For comparative purposes, it is convenient to consider the capacity of a system per unit of drawdown. This parameter is known as specific capacity, and for the purposes of this study is defined as the discharge rate of a well divided by the driving head (or head loss between the recharging river and the well), with units of cubic meter per day per m (m^3/d/m, or m^2/d, or mgd/ft) drawdown:

System Specific Capacity = (pump discharge)/(river water level – well water level)

In systems where nearby wells impact the water level in the aquifer, the driving head will be reduced. For the system at Louisville, no interfering wells exist, and the driving head is simply the river level minus the water level in the caisson.

Specific capacity at a given RBF site can vary with time as a function of changes in riverbed conductivity, aquifer conductivity (bioclogging and geochemical clogging), wellscreen deterioration, and temperature. Temperature effects are primarily equal to viscosity changes, but temperature may also impact bioclogging and geochemical clogging. Temperature alone accounted for the doubling of specific capacity between winter and summer at one of the sites evaluated in this research.

Traditionally, specific capacity has been used to evaluate the performance of individual wells only. However, in the context of this research, specific capacity was also used to evaluate the performance of the entire wellfield collectively. The term "system specific capacity", when used is this study, refers to the total discharge from the wellfield and the average drawdown in the wells.

CHANGES IN SPECIFIC CAPACITY WITH TIME AT LOUISVILLE

Figure 6.1 provides the calculated values for specific capacity from the beginning of pumping through 2004 for the well at Louisville. Data were unavailable for the period of February 2002-October 2002 due to a loss of computer-generated data. From November 16, 2003 through April 9, 2004 flow from the well was reduced from 64,000 m^3/d to 38,000 m^3/d (17 to 10 mgd) as the result of a pump failure, coinciding with an apparent increase in specific capacity. The well was out of service for repair from April 15 through July 28, 2004. On April 15 2004 the river experienced a sharp flood wave, cresting 2.5 m above pool on April 19 and returning to normal pool on April 23. Upon re-start of the well July 28, 2004 at 64,000 m^3/d (17 mgd) another increase in specific capacity was observed. (See Figure 6.1).

The higher values of specific capacity observed from August 2004-December 2004 are likely the result of riverbed surface renewal during the flood event when the well pumps were off. The aquifer was charged with colder water prior to well shut-down in April 2004, which

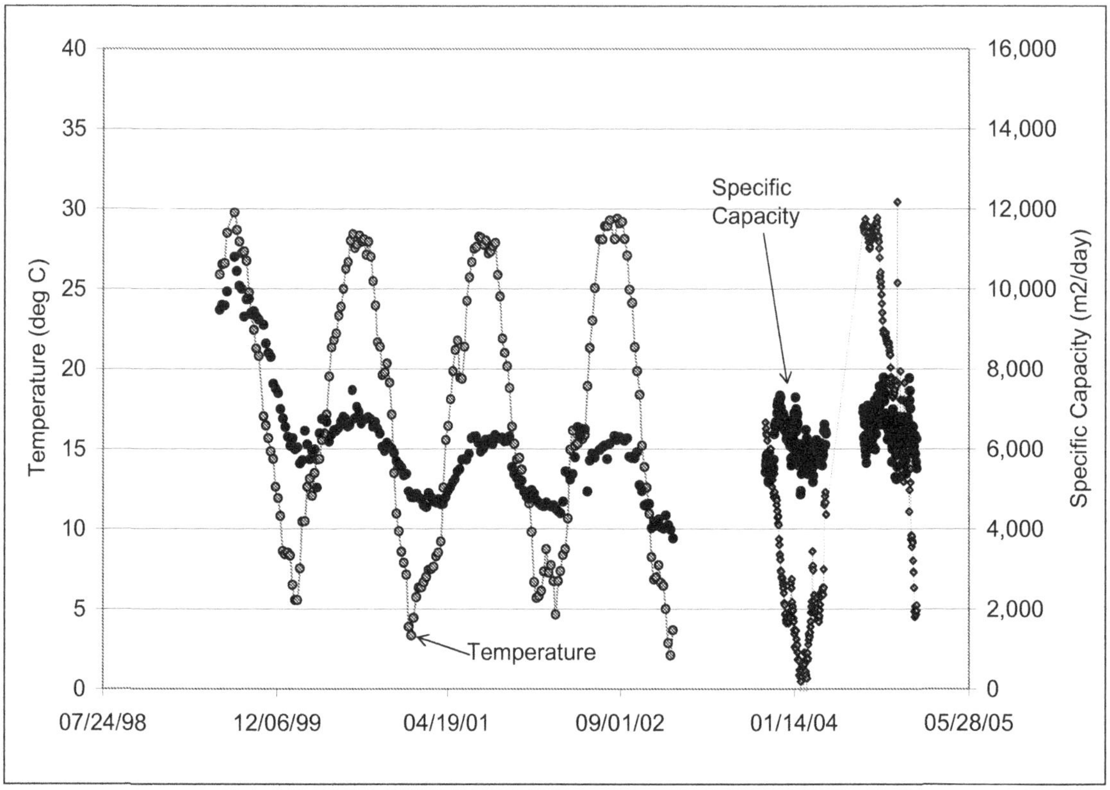

Figure 6.1 Variation of specific capacity with time and river temperature

would have been expected to have a negative impact on specific capacity values upon start up. Specific capacity for this system appears to have stabilized between 5000 and 7000 m^2/d (0.4 to 0.56 mgd/ft) at a typical flow of 64,000 m^3/d (17 mgd).

The annual decrease in specific capacity observed during the first 4 years of operation was assumed to be a function of riverbed clogging with sediment from infiltrated water. This annual decrease was well represented statistically as function of the natural log of time from initiation of pumping, with the magnitude of the decrease in specific capacity diminishing with time (Hubbs, 2004). The impact of streambed scour as a restorative agent was observed, and the impacts of decreased riverbed hydraulic conductivity with time and restorative impacts of scour appeared to balance after about 4 years.

Data from Figure 6.1 were averaged across 12 month periods from July 1999 through July 2003. The decrease in the maximum, minimum, and average specific capacity for these 12-month periods is presented in Table 6.1. The high level of reduction in the first year is skewed by the high water temperature in both the river and the aquifer, as the aquifer was charged with warm water, replacing groundwater at a temperature of 15°C. For all subsequent years, the aquifer temperature reflected recharge from the coldest to warmest seasons, with a minimum river recharge temperature as low as 3 °C. It is difficult from these data to determine how much of the decrease in specific capacity is the result of start-up temperature effects, and how much is a result of riverbed clogging in the first year of operation.

Table 6.1: The reduction of specific capacity across 12 month periods, 7/99 through 7/03

	Specific Capacity: m^3/day/m			Specific Conductance Compared to 1999-2000 (percentage of)		
Year	**Annual Average**	**Annual Minimum**	**Annual Maximum**	**Annual Average**	**Annual Minimum**	**Annual Maximum**
1999-2000	7751	5950	10625	100%	100%	100%
2000-2001	5694	4693	6849	73%	79%	64%
2001-2002	5519	4502	6287	71%	76%	59%
2002-2003	5021	3765	6280	65%	63%	59%

SPECIFIC CAPACITY VERSUS TEMPERATURE

The temperature of the water in the river and in the aquifer determines the viscosity of water as it passes from the river through the riverbed and aquifer to the well. Data presented in the previous section demonstrated the strong correlation between temperature and specific capacity at the study location at Louisville. A more detailed statistical analysis of specific capacity as a function of river temperature, well temperature, and time was conducted on the dataset from LWC for the years 1999 through 2003 (Hubbs 2004). Noting that data were available from the initial start-up of this system, time was used as a surrogate for time-dependent clogging. Using multilinear regression techniques for variables of river temperature, well temperature, and the log of time, a multiple R regression value of 0.96 was obtained. Equation coefficients from this analysis are presented in Figure 6.2. Excluding the time surrogate for clogging, the multiple R regression value was 0.62. This indicated that for the initial start-up of this system, time-dependent clogging was a significant factor in the variation of specific capacity along with temperature.

The significance of clogging during the first few years of operation of the system at Louisville was further supported by computer-model estimations of riverbed leakance[2] conducted between 1999 and 2003 (Schafer, 2006). These data indicate a rapid decrease in riverbed leakance with time between 1999 and 2001. These data are presented in Figure 6.3.

It is noted that the regression analysis used time as a surrogate for clogging, and that the computer modeling of leakance assumes that all other wellfield characteristics are known. Thus, neither technique provides a direct measure of clogging in the riverbed as a result of pumping. However, the data from both techniques are supportive of the conceptual model of time-dependent clogging balanced against riverbed scouring.

[2] Leakance is the riverbed hydraulic conductivity divided by the riverbed thickness, thus yielding a parameter with units of inverse time.

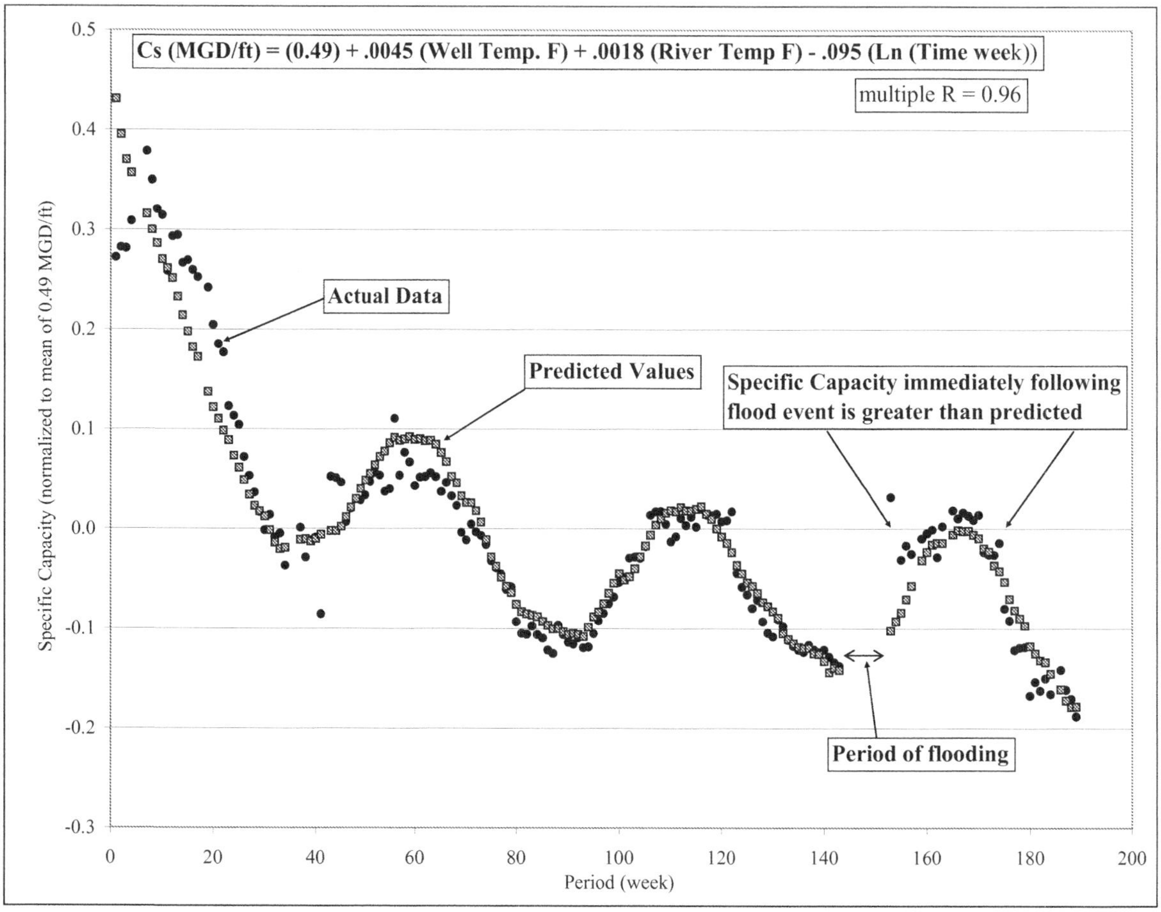

Figure 6.2 Specific capacity as a function of time and temperature

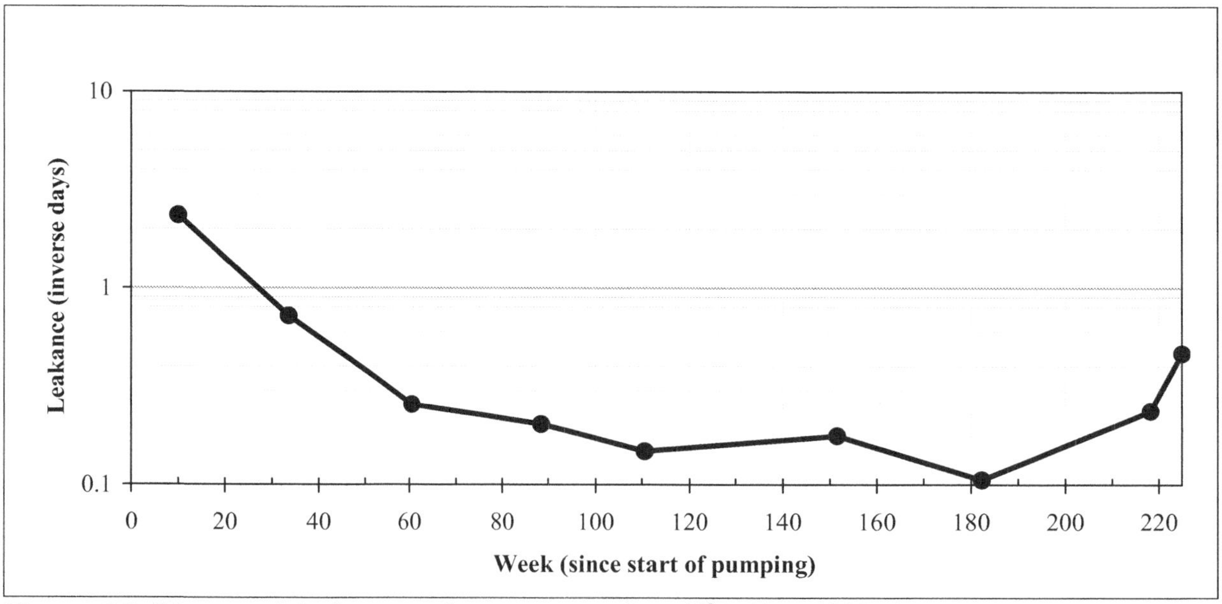

Figure 6.3 Measured leakance values corrected to 68°F (Oct 1999 river temperature)

Water utilities at three other study locations provided adequate data necessary to perform a regression analysis on temperature and specific capacity. The data needed for the analysis included river level, well water level, production rates, and temperature data for the aquifer and/or river. In the case where both well and river temperature data were available, the analysis was run twice; once using the well temperature data, then again using the river temperature data. The results of the analysis that resulted in the highest regression coefficient (R^2) are presented.

The Des Moines Water Works provided data for the period from October 2001 to December 2002 for the five radial collector wells at the study site adjacent to the Raccoon River in Des Moines, Iowa. The data were taken less frequently than those from Louisville. Data were not provided for the horizontal well. Drawdown and production data were available for each of the five radial collector wells. The total production and average drawdown data for the five wells were used to calculate a system specific capacity for the entire wellfield. The river temperature varied between 2 and 29 degrees Celsius. The system specific capacity ranged from 10,000 to 27,000 m^2/d (0.8 to 2.2 mgd/ft). A strong correlation between specific capacity and river temperature was evident even though the dataset was small. A regression analysis between bulk specific capacity and river temperature yielded a regression coefficient (R^2) of 0.7. Specific capacity and temperature data for the Raccoon River are presented in Figure 6.4.

The City of Lincoln provided data for the two radial collector wells that are located on an island in the Platte River in Ashland, Nebraska. Data for the first well, Well 90-1, consisted of 18 data points for the period from September 2003 to September 2004. Data for the second well, Well 90-2, consisted of 12 data points for the period from December 2003 to September 2004. The influence of well temperature on specific capacity was more significant than the influence of river temperature. A regression analysis using the data from Well 90-1 revealed a significant correlation between specific capacity and temperature with a regression coefficient of 0.54. However, a similar analysis for Well 90-2 yielded a regression coefficient near zero. The poor correlation between specific capacity and temperature at Well 90-2 might have been the result of the limited dataset and/or the influence of pumping from nearby wells. Specific capacity and temperature data for Well 90-1 are presented in Figure 6.5.

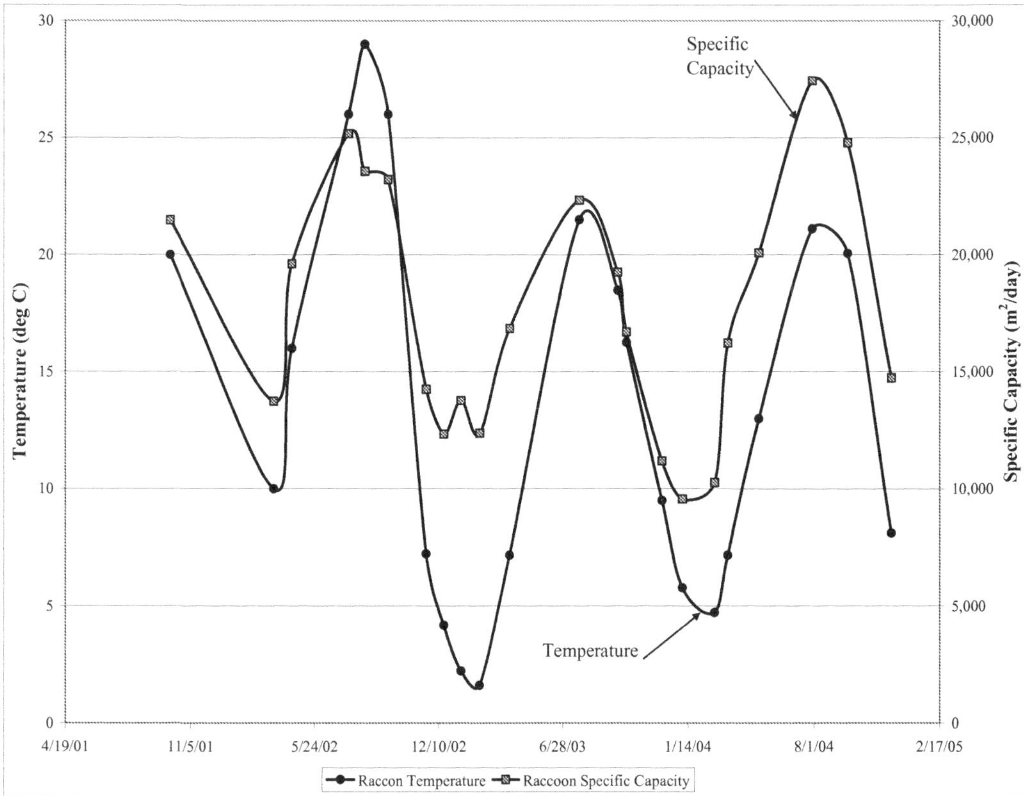

Figure 6.4 Raccoon River specific capacity and temperature

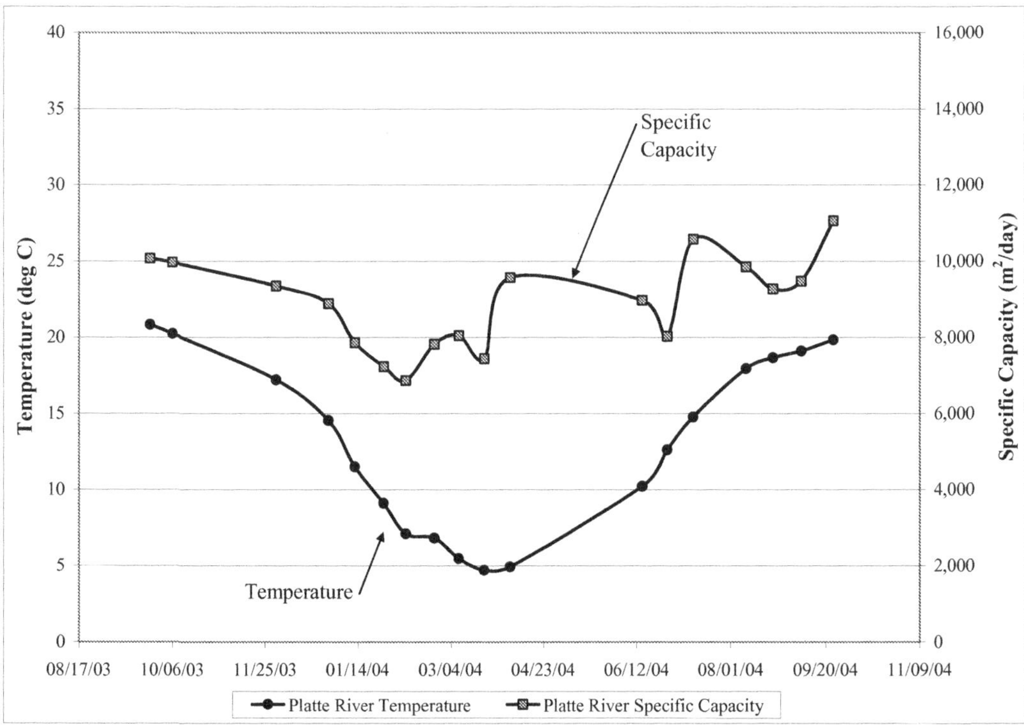

Figure 6.5 Platte River specific capacity and temperature

The Kansas City Board of Public Utilities (KC BPU) provided data for the radial collector well adjacent to the Missouri River in Kansas City, Missouri. The dataset consisted of 29 data points for the period from October 2002 to November 2004. Well temperatures varied between 9 and 26 degrees Celsius while river temperatures varied from 0 to 29 degrees Celsius. The specific capacity ranged from 15,000 to 26,000 m^2/d (1.2 to 2 mgd/ft). The influence of well temperature on specific capacity was more significant than the influence of river temperature. A regression analysis of the data produced a correlation coefficient of 0.81 between well temperature and specific capacity. Specific capacity and well temperature data for the study site in Kansas City, Missouri is presented in Figure 6.6.

The effect of temperature variation on specific capacity was evident for all four systems for which data needed to perform the analysis were available. The specific capacity varied in cycles that mimicked the annual temperature variations. The specific capacity was plotted versus temperature as shown in Figure 6.7 through Figure 6.10. Figure 6.7 shows the data for the specific capacity versus temperature and trend lines for the four study sites. Figure 6.8, Figure 6.9, and Figure 6.10 show normalized specific capacity and temperature. The specific capacities were normalized by dividing all specific capacity values for each system by the highest specific capacity value of that system. Figure 6.8 shows the normalized specific capacity versus temperature and trend lines for the four systems analyzed. Figure 6.9 is intended to show the overall trend of temperature and normalized specific capacity. However, because data from the Ohio River included warm-weather data from the initial start-up of the well, the Ohio River data did not match well with the other systems and is separated from the other data points.

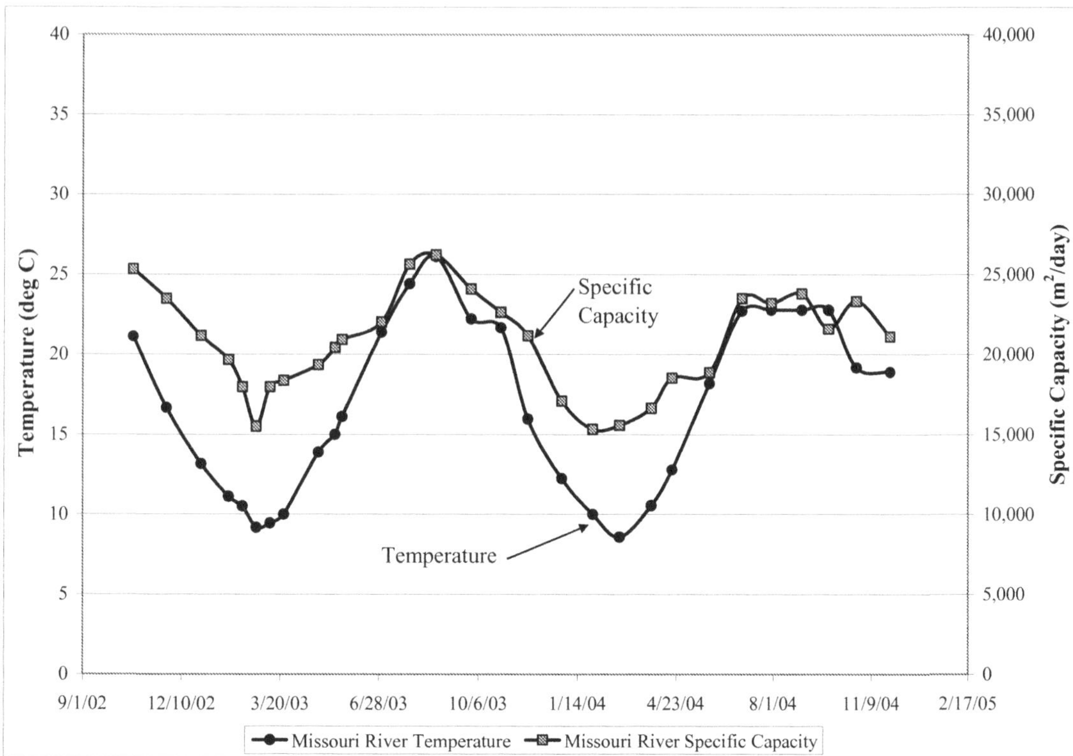

Figure 6.6 Missouri River specific capacity and temperature

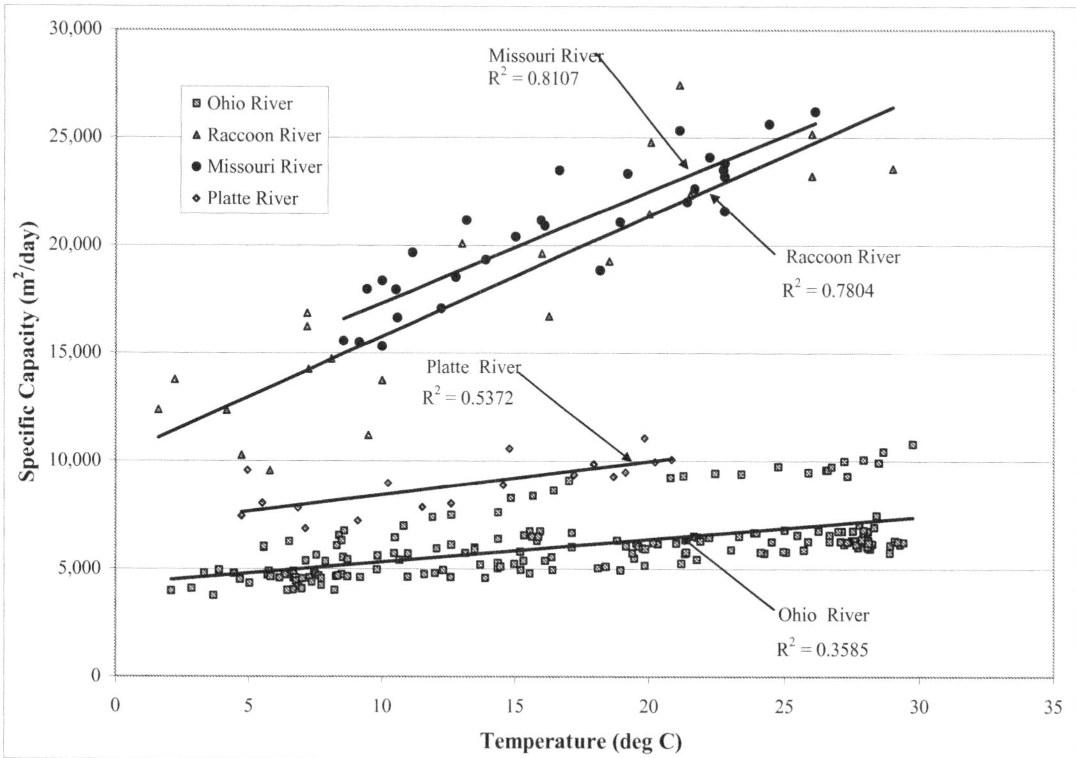

Figure 6.7 Specific capacity and temperature by study site

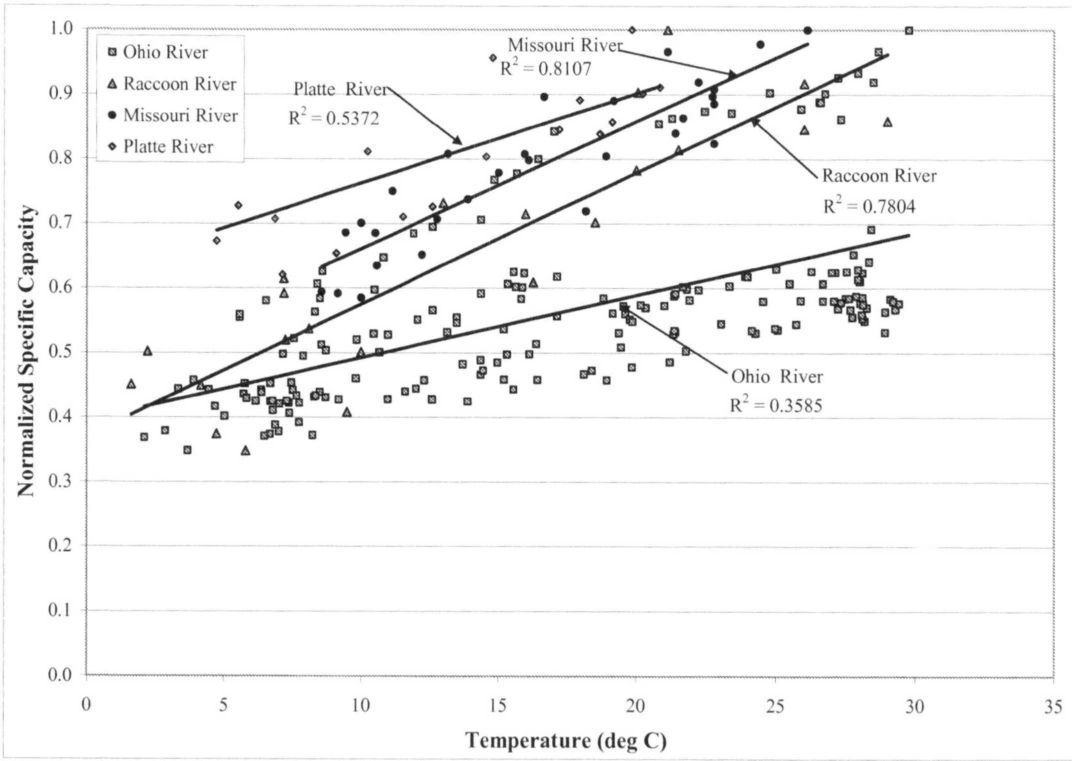

Figure 6.8 Normalized specific capacity and temperature by study site

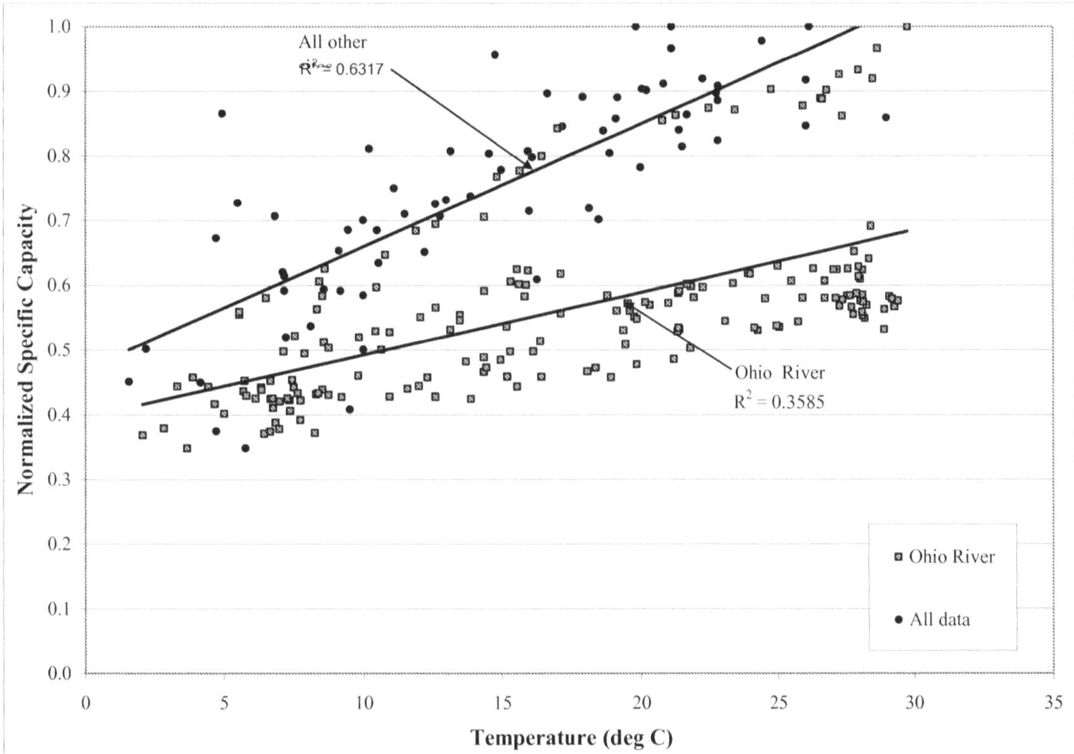

Figure 6.9 Trend of normalized specific capacity and temperature, showing data from the Ohio River as an outlier.

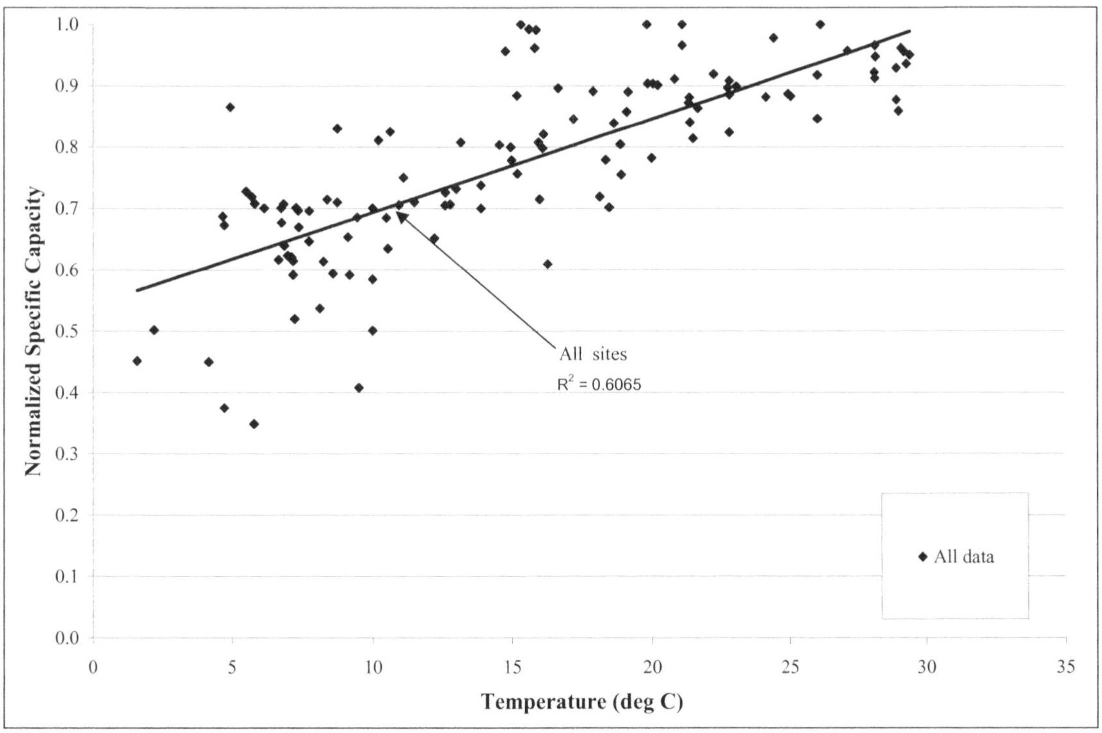

Figure 6.10 Overall trend of normalized specific capacity and temperature, using only data from 2002 for the Ohio River dataset.

Figure 6.9 is much like Figure 6.10, except that only data from the 2002 year for the Ohio River site was utilized and combined with data from the three others sites to generate an overall trend of temperature and specific capacity. This was done to lessen the influence of the time-dependent decrease in specific capacity from the first years of pumping at the Ohio River site.

As seen in Figure 6.8, the normalized specific capacity exhibited a nearly linear relationship with temperature. All four systems showed correlations between river or well temperature and specific capacity. The analysis of data from the Ohio River site resulted in a lower regression coefficient than the other sites because of the decrease in specific capacity with time during the first two years of pumping.

The specific capacities of the systems on the Ohio, Raccoon, Platte, and Missouri Rivers are shown for the minimum, average, and maximum temperatures in Table 6.2. All the systems analyzed showed a significant variation in specific capacity with temperature. The specific capacity of the system on the Raccoon River more than doubled as the temperature cycled from its minimum to maximum values. This doubling demonstrates the importance of considering the water temperature when evaluating a possible RBF location and choosing the dates of a pump test. A test performed when the infiltrating water temperature is highest could significantly overestimate the average yield for a proposed system.

The hydraulic conductivity of the riverbed and aquifer should vary inversely with the viscosity of the water. An attempt was made to normalize the data in Table 6.2 by multiplying specific capacity and the viscosity value that corresponds to the temperature at which the specific capacity was measured. Viscosity values were taken from Davis and Cornwell (1998). The viscosity values and the products of specific capacity and viscosity are shown in Table 6.3.

The variations of the specific capacity values in Table 6.2 were compared with variations of normalized specific capacities values in Table 6.3. The normalized specific capacities for the Raccoon River, Platte River, and Missouri River sites exhibited significantly less variation than the corresponding specific capacity values in Table 6.2. This supports the conclusion that most

Table 6.2: Specific capacity and temperature summary

	Temperature (deg C)				Specific Capacity		
River	Min	Avg.	Max	Location	Min .	Avg.	Max
Ohio	2	17	30	River	4000	5000	6000
Raccoon	2	13	29	River	10000	18000	27000
Platte	5	13	21	Well	7000	8000	11000
Missouri	9	17	26	Well	15000	24000	26000

Table 6.3: Specific capacity and viscosity

	Viscosity (cP)				Specific Capacity*Viscosity			Time Lag (days)
River	Min.	Avg.	Max.	Location	Min.	Avg.	Max.	
Ohio	1.67	1.08	0.80	River	6680	5410	4806	40
Raccoon	1.67	1.27	0.82	River	16700	22860	22086	
Platte	1.52	1.27	1.00	Well	10640	10160	11000	30
Missouri	1.35	1.08	0.87	Well	20250	25968	22698	25

of the variation in specific capacity at these sites can be explained by the change in viscosity due to temperature changes. The normalized specific capacity on the Ohio River exhibited nearly the same magnitude of variation (but reversed) as the original specific capacity values. This might be explained by the loss of specific capacity with time at the Ohio River site and the time lag between the well and river temperature at that site. As shown in Table 6.3, the time lag for the Ohio River was approximately 40 days.

OPERATIONAL FIELD DATA FROM PARTICIPATION SITES

The operational data from the participating RBF sites were split into two tables. The production and drawdown characteristics are presented in Table 6.4. The temperature characteristics are found in Table 6.5. The sustainable capacity was estimated by multiplying the average specific capacity by the available drawdown. The Platte River had the highest estimated sustainable specific capacity.

There was a clear delineation between the available drawdown reported for study sites in the United States and study sites in Europe. The available drawdown at European sites ranged from 0 to 8 m. The available drawdown for the United States sites ranged from 14 to 19 m. Since the aquifers in the United States and Europe were comparable, the difference in reported available drawdown can be attributed to differences of judgments of personnel at utilities in the United States and Europe about the amount of drawdown available for a given situation. Utilities in Europe were more conservative and less likely to fully utilize the aquifer than utilities in the United States.

Table 6.4 Operational characteristics –1

River	Wellfield	Average Pumping Rate (m^{3}/**day**)	Average Drawdown in Well (m)	Sys Spc Capacity (m^2/day)	Available Drawdown (m)	Maximum Capacity (m^3/**day**)
Llobregat	**Cornellà**	**62,000**	**2.0**	**31,000**	**0 ***	**0 †**
Rhine	**Grind**	**106,000**	**4.0**	**26,500**	**8**	**212,000**
Rhine	**Flehe, PW V**	**33,000**	**1.7**	**19,000**	**5**	**97,000**
Rhine	**Staad, PW I**	**25,000**	**2.0**	**12,500**	**5**	**63,000**
Enns	**KRB02**	**2,000**	**0.8**	**2,160**		
Ohio	**BE Payne**	**64,000**	**11.0**	**6,000**	**19**	**110,000**
Raccoon	**Maffitt**	**45,000**	**2.4**	**19,000**	**19**	**360,000**
Missouri	**Nearman**	**125,000**	**6.1**	**20,000**	**14**	**290,000**
Great Miami	**Bolton**	**60,000**				
Russian River	**Wohler / Mirabel**	**142,000 ‡**				
Platte	**Ashland**	**76,000**	**3.0**	**26,000**	**19**	**480,000**

*** The water level in the aquifer is below sea level. The global abstraction from the system is causing severe saline intrusion (Martin-Alonso., 2005).**

† Current production practices overexploit the aquifer and are unsustainable (Martin-Alonso 2005).

‡ Data given are the average production in the river when the inflatable dam is not raised and diversion ponds are not filled.

Table 6.5 Operational characteristics –2

River	Wellfield	River Temperature for a Typical Year (deg C)			Well Temperature for a Typical Year (deg C)			Temperature * Lag between River and Well (days)
		Min.	Avg.	Max.	Min.	Avg.	Max.	
Llobregat	Cornellà	5	18	30	11	17	24	N/A †
Rhine	Grind	4	14	26	11	14	17	90
Rhine	Flehe, PW V	4	14	26	10	14	18	45
Rhine	Staad, PW I	4	14	26	10	14	18	35
Enns	KRB02	1		16				4
Ohio	BE Payne	3	17	29	12	18	26	60
Raccoon	Maffitt	1	14	25				
Missouri	Nearman	1	15	29	9	18	26	25
Great Miami	Bolton	1	16	31				
Russian River	Wohler / Mirabel							
Platte	Ashland	0	15	28	5	13	22	30

* Temperature lag was graphically estimated by visual comparison of time-series graphs of river and well temperature.

† Data given on transient time between the scarification area and the abstraction area under normal conditions show the transient time to be approximately 1460 days. The transient time differs when large amounts of water infiltrates from the river during periods of flooding. (Martin-Alonso 2005)

CHAPTER 7
SUMMARY AND CONCLUSIONS

SUMMARY

Data compiled from 11 RBF sites and the in-depth site specific data from Louisville provide a background against which other potential RBF sites can be evaluated. Typical ranges for river hydrology parameters, riverbed composition, aquifer characteristics, and wellfield operating data are provided in a format that allows quick comparison to other sites. The basic elements of stream hydrology, riverbed dynamics, and aquifer characteristics are introduced with each chapter to provide the background against which the technology of RBF is presented. Detailed information from the RBF facility at Louisville provides specific data on stream hydrology, riverbed hydraulic conductivity, aquifer transmissivity, and the influence of temperature and riverbed hydraulic conductivity on wellfield specific capacity.

The impact of unsaturated conditions developing in an aquifer under the riverbed is presented, and the static and dynamic forces acting on the riverbed are considered. A hypothesis is offered to explain the dramatic decreases observed in riverbed hydraulic conductivity as a result of clogging and the dewatering of the aquifer under the riverbed.

This study demonstrates the importance of river hydrology and the riverbed/aquifer interaction in the capacity of RBF systems. The report provides basic information to assist those interested in RBF with regards to assessing the adequacy of sites for RBF, and considering the options with regards to types of wells and collection systems. Recommendations are developed to assist in planning for a project involving RBF.

CONCLUSIONS

Data collected from 11 RBF systems in the United States and Europe and the site-specific data from Louisville yield the following conclusions:

1. RBF has been implemented successfully in a variety of complex conditions in the United States and Europe. The data compiled in this study provide a background against which future RBF sites can be compared.
2. The riverbed/aquifer interaction is critical in the recharge and long-term capacity of RBF systems. This interaction should be considered in the design of RBF systems.
3. Riverbed clogging in RBF systems can result from mechanical, biological, and geochemical processes.
4. The impact of unsaturated conditions under the riverbed on riverbed hydraulic conductivity and well specific capacity is dramatic, resulting in a decrease in specific capacity and the extension of the point of recharge further from the well.
5. The impact of temperature on yield from RBF systems is dramatic because of the wide range of water temperature experienced at RBF sites. A strong correlation was developed between specific capacity and temperature at individual sites, as shown by data from four RBF systems. Specific capacity of an RBF system can more than double as temperatures in the aquifer and river fluctuate from annual lows to annual highs.

6. Data from the 11 sites were inadequate to quantify the effects of other parameters such as river discharge and river water quality on the capacity of RBF systems. This uncertainty was likely due to limitation in the number of sites studied.
7. Data from Louisville showed that specific capacity can decrease significantly in early stages of system development.

FUTURE RESEARCH

While it is known that scour is important to the renovation of riverbed surfaces in RBF systems, the intensity and depth of beneficial scouring is site specific, with little data available in the literature. Additional research into the depth of clogging and required depth of scour to yield beneficial riverbed renewal is warranted.

The compilation of basic operational data from RBF facilities will lend increasing insight into the significance of riverbed and aquifer characteristics in RBF design. Specifically, temperature data between the river and the well discharge reflect a lag in peaks, a shift in averages, and a magnitude difference between maximum-to-minimum values. These three parameters can yield valuable information regarding riverbed and aquifer hydraulic conductance and groundwater contribution to the well discharge. These operational data should be recorded and analyzed for trends.

This research has indicated that relatively large hydraulic forces can be exerted on the riverbed when the media below becomes unsaturated. This phenomenon should be further explored at the laboratory level, evaluating the hydraulic conductivity and shear resistance before and after exposure to unsaturated conditions.

CHAPTER 8
RECOMMENDATIONS ON DESIGN CONSIDERATIONS

This chapter is written to provide insight to utilities regarding the type of information needed when considering the design of an RBF system. While it is not to be considered exhaustive, it represents logical steps in planning for an RBF system.

SITE-SPECIFIC LITERATURE REVIEW

Much information is available in published literature for many of the river valleys suited for RBF. In some instances, there may already be large capacity systems serving industrial customers nearby, with unpublished operating data that can be invaluable to assessing aquifer conditions. Gravel companies likewise can be a valuable source of unpublished information on composition of riverbed media. All available data should be compiled and reviewed as a first step in assessing a site for RBF.

AQUIFER TESTING

Traditional aquifer testing remains a critical element in any well system design. Proper layout of the pumping well and piezometers is necessary to accurately estimate aquifer transmissivity and the impact of the river recharge on the pump test results. Wellscreens should either be fully penetrating, or in the case of piezometers, nested to allow proper characterization of the aquifer. Aquifer media should be sampled in a way to preserve the fines in the sample for particle size analysis, and the entire depth of aquifer should be sampled at appropriate intervals. If boulders are likely in the aquifer and a horizontal collector well is being considered, the drilling and sampling technique should be adequate to indicate the presence of boulders and clearly identify the bedrock elevation.

If feasible, piezometers should be placed in the stream itself. In sand-bottom streams, probes for pressure and temperature can be installed by jetting a hole in the riverbed, placing the probe to the desired depth, and jetting a trench to the shore for the wire.

Piezometer layout should consider the need to analyze aquifer parameters during the life of the RBF system, and should compliment a long-term monitoring program designed to differentiate between hydraulic conductivity changes in the riverbed, aquifer, and at entrance to the wellscreens.

The conditions under which the pump test is run can be critical to data interpretation, and should be carefully noted. The test should be supervised by a qualified hydrologist. Data collected for aquifer characteristics can be compared to data in Table 5.1 for other RBF sites.

STREAM HYDROLOGY

The recharging stream should be characterized by stage and flow hydrographs, noting whether the stream is impounded at any time of the year. The stream slope should be computed, and if available, stream velocity profiles should be analyzed to estimate streambed shear stress at various flows. The physical location of the site with respect to river bends, obstructions, dams, side streams, and bathemetry should be analyzed with respect to these attributes impacting riverbed shear stress, scour and deposition patterns. During high flow events, the site should be visually observed if possible, looking for evidence of flow disturbances such as flow separations (upstream flow near the bank). These data should be compared to the data in Table 3.1 and Figure 3.6.

RIVERBED SEDIMENT ANALYSIS

A literature search on riverbed sediment transport for the site under consideration should be conducted. Data will most likely be found in USGS and highway design publications, dating back to the 1970s. Information should be sought from secondary sources (gravel operators) with regards to dredging history and records. Riverbed samples should be collected from the site(s) being considered, preferably by diver, looking for indications of media size, armor coats, debris, scour, and sediment transport. Riverbed samples should be analyzed for particle size distribution, and the riverbed should be viewed macroscopically for indications of scour intensity and frequency. These data should be compared to the information in Table 4.4 and Appendix B.

STREAM WATER QUALITY

The water quality of the stream should be considered from its potential to support excessive microbial growth resulting in bioclogging. Streams with substantial upstream wastewater discharges should consider the impact of stream quality on bioclogging. Data on stream quality should be compared to data in Table 5.2, Table 5.3, and Table 5.4 for comparison to other RBF sites.

GROUNDWATER WATER QUALITY

Groundwater samples from the site should be collected during the aquifer pump tests and analyzed for chemicals listed in Table 5.4. These data should be reviewed by a qualified water quality expert to determine any tendency towards bioclogging and geochemical clogging, in the aquifer and at the screens. Depending on the mix of surface water and groundwater in the well discharge, the quality of the well discharge from a fully operational RBF system will reflect partial characteristics of both the stream and the groundwater.

ESTIMATING SUSTAINABLE SPECIFIC CAPACITY

Initial estimates of specific capacity can be made by modeling the aquifer and proposed RBF system, given aquifer characteristics and estimates for riverbed conductivity. This initial estimate should be considered in light of the seasonal temperature ranges to be expected during operation, and any decrease in riverbed conductivity that might be anticipated during the first several years of pumping, either through particle clogging, bioclogging, or geochemical clogging of the riverbed.

If site conditions allow, an excellent option is to construct the test well for the initial aquifer characterization in a way that allows the discharge to be utilized in the system, either as influent to the treatment plant or for discharge to the distribution system after appropriate treatment. This allows for additional long-term data collection to help define both hydraulic and water quality questions for the proposed system. If possible, such systems should be allowed to run for at least a year, and preferably longer.

ONGOING HYDRAULIC MONITORING

After the system is operational, routine (or automated) monitoring should be conducted for river level, well level, river temperature, well temperature, and well discharge. These data will allow the specific capacity of the system to be tracked and the temperature influences to be analyzed. If automated, data should be collected and stored in a manner to allow trends in specific capacity to be followed. This might include daily average readings or weekly average readings. If discrete data are used for statistical analysis, care should be taken to collect data at a time when the system is most stable with relation to well discharge flow rate.

Aquifer water level readings from piezometers should be taken periodically (4 times per year), during a time when river level and well flow is relatively stable. These readings might also be taken following a flood event to provide insight to the impact of riverbed scouring on riverbed hydraulic conductivity.

Initial specific capacity calculations can be expected to be influenced by the temperature dynamics between the river and the aquifer, with warm-river water start-ups providing the conditions for highest initial specific capacity. Temperature variations in the aquifer tend to stabilize after one annual cycle, providing site-specific data on the range of specific capacity values to be expected across the annual temperature variation. At Louisville, the range of impact on specific capacity was from 4,000 to 6000 $m^3/m/d$ (0.4 to 0.56 mgd/ft) across a temperature swing of 25^0C in the river and 15^0C in the well discharge after 3 years of operation.

A gradual decrease in specific capacity can be expected as the piezometric surface in the aquifer balances between well discharge and aquifer recharge. The specific capacity might also be expected to decrease if the piezometric surface extends into the river and unsaturated conditions develop under the river, as the effective point of recharge extends further into the riverbed and/or upstream and downstream. At Louisville, the decrease in annual average specific capacity from 1999 to 2002 was to approximately 65% of the originally calculated value. Because pumping started in the summer and charged the entire aquifer with warmer water, the variation between specific capacity from the annual 1999 dataset and the remaining years cannot be attributed solely to decreases in riverbed hydraulic conductivity; at least a portion of this must be attributed to the temperature dynamics at start-up. However, the variations between the 2000 dataset and the 2002 dataset have little influence from start-up anomalies, and the bulk of the decrease can be attributed to reduction in riverbed hydraulic conductivity. Between July 2000 and July 2002, the annual specific conductivity decreased to 88% of the July 2000 value.

Frequent bed scour capable of moving the bulk of the bed material will result in less reduction in riverbed hydraulic conductivity with time. Ongoing analysis of flow hydrographs, scour shear stresses, and riverbed material size can provide insight into changes in riverbed hydraulic conductivity that is reflected in the specific capacity of the wellfield.

WELLSCREEN OPTIONS FOR WATER EXTRACTION

Most RBF systems have been designed as either vertical wells or horizontal radial collector wells with caissons. Horizontally drilled wells have been constructed, and some systems are constructed as infiltration galleries. The choice of extraction system is influenced by the site hydrology and site-specific concerns with constructability and physical appearance. The following comments are limited to the interaction of riverbed hydraulic conductivity and aquifer conductivity in the selection of a specific type of well.

Two hypothetical situations are presented to illustrate the impact of extraction method on long-term specific capacity. The most ideal situation for minimizing riverbed clogging in an RBF system would be to have a continuous wellscreen directly under the riverbed in a deep aquifer, with aquifer media well sorted from most course to least course from the screen to the riverbed. Such a system distributes the flow evenly across the entire riverbed, with flux rates being low and equal for a given rate of well discharge. Given adequate riverbed shear stress, the entire riverbed can be renewed if the extraction rate were appropriate for the length of riverbed impacted.

The least ideal situation would be a fully penetrating single vertical well (or widely spaced series of single wells) very near the river's edge in a completely homogeneous sand bed, with a very steep radial drawdown curve to the river. This system results in flow concentrated towards the point of extraction, with flux rates highest near the well, decreasing as a function of distance from the well, for a given rate of well discharge. This system is at greater risk of clogging near the well due to the increased flux rates, and may also result in the aquifer becoming unsaturated under the riverbed near the well, further exacerbating the clogging. This system places uneven stresses on the riverbed, and riverbed shear stresses may be inadequate to renew the riverbed at points where the stress from clogging and loss of saturation are greatest.

In both situations, the rate of extraction determines if riverbed clogging will occur. If extraction rates are low enough, clogging will occur in neither case; and if the extraction rate is extreme, clogging will occur in both cases. Between these two extremes, the greater specific capacity will result from the most ideal situation, where the flux rates and stresses are distributed and minimized across the riverbed.

The most ideal situation is likely unfeasible from a construction perspective, and the least ideal is likely the most economical. Between these two extremes, however, some general trends can be noted:

1. Shallow aquifers with moderately conductive lower layers may be better suited to radial collector wells, where a greater length of wellscreen can be used to exploit the conductive lower layer of aquifer.
2. Deeper aquifers with highly conductive lower layers may be better suited to vertical wells, where the lower layers act as underdrains to the vertical wellscreen set deep in the aquifer, and the length of wellscreen is adequate for the desired capacity.

The selection of water extraction technique in a given system will likely be influenced by factors of aesthetics and economics as well as hydraulics.

Most aquifers are neither homogeneous nor ideally sorted, and the media size and conductivity in most aquifers varies to some extent in every direction. Many glacially formed river systems have ancient riverbeds with media sorted generally from larger to smaller from the bottom of the ancient river channel to the top of the aquifer. In an ideal situation, a thick, well sorted gravel bed will underlie a thick sand bed in the aquifer, acting as an underdrain to recharge from the hydraulically connected river. In such systems, the extraction technique (vertical versus horizontal) is less critical, given adequate length of wetted wellscreen in the formation. Because of the high conductivity of the underlying layer of gravel, the head loss in this part of the formation will be minimal, allowing water to be extracted from a relatively long stretch of adjoining river. This condition minimizes the flux rate across the riverbed near the well, and thus decreases the tendency towards clogging.

COLLECTION TECHNIQUES

The method of collecting water from a series of wells in an RBF system has little impact on the riverbed hydraulics, except for implications on how they allow water to be extracted from the RBF system. Those collection techniques that accommodate extraction systems providing the most equal distribution of flux rates across the aquifer and riverbank would be integral in an overall system designed to minimize clogging and maximize sustainable yield.

Three basic collection techniques have been considered for RBF systems:

- conventional systems with pumps in each well discharging to a common header.
- siphon systems with a collection siphon attached to a series of wells and ending in a common pump station (Schubert 2002b)
- gravity tunnel systems with a series of wells connected to a tunnel or pipe in or below the aquifer and connected through gravity flow to a common pump station (Hubbs 2002a).

Each collection system has unique advantages impacting the hydraulic, economic, and aesthetic elements of an RBF system. In general, system designs that most closely approach a continuous extraction system along a given stretch of riverbank will result in the least impact in riverbed clogging and specific capacity deterioration with time.

APPENDIX A
DATASOURCES FOR PARTICIPATING SITES

This section describes the primary sources of data for each location. It is worth noting that data taken from the USGS website (http://waterdata.usgs.gov/nwis) was used extensively for all study sites in the United States. The USGS gages were searched by stream name to determine the number and types of relevant gages. Where available, river hydrology, river geometry, water quality, bed sediment, and suspended sediment data were taken from gage locations near study sites. Other gages located along the lengths of the streams were used in the determination of water surface elevations for the generation of the water surface profiles found in Figure 3.6 and 3.7. The exception was the Ohio River profile, which was generated based on the navigation profile from the United States Army Corps of Engineers (USACE).

Ohio River at Louisville, Kentucky. Data for the Ohio River was obtained primarily from the Louisville Water Company (LWC), USGS, Ohio River Valley Water Sanitation Commission (ORSANCO), and publications related to studies performed at the site.

LWC provided much of the operational data, site characteristics, and water quality data needed. USGS data for gage 03294500 at Louisville, Kentucky was used to determine the hydrologic characteristics of the site. ORSANCO provided additional water quality data (Martin 2005, Yongstrom 2005). Much of the aquifer characteristics were found in published results of studies performed at the site. Additional data were acquired from the United States Army Corps of Engineers (USACE) and other sources.

Great Miami River near Cincinnati, Ohio. The primary sources of data for the Great Miami River near Cincinnati, Ohio included: USGS, Greater Cincinnati Water Works (GCWW), and publications related to studies performed at the wellfield. Most of the hydrology information for the site came from USGS gage USGS 03274000 – Great Miami River at Hamilton, Ohio.

Raccoon River at Des Moines, Iowa. The primary sources of information for the study site in Des Moines, Iowa were the Des Moines Water Works and USGS data. USGS data was obtained for gage 05484900 – Raccoon River at Fleur Drive in Des Moines, Iowa (hereafter referred to as Fleur Drive gage) when available. However, some data needed for the study were not available for the Fleur Drive gage. Data from USGS gage 5484500 – Raccoon River at Van Meter (hereafter referred to Van Meter gage), approximately 50 km upstream, was used to supplement the data from the Fleur Drive gage when necessary. Production data were acquired for the five radial collector wells only. Data for the inclined horizontal well were not available within the timeframe of the study.

Platte River near Ashland, Nebraska. Data for the study site near Ashland, Nebraska were obtained from the City of Lincoln and USGS for gages 06801000 and 06805500 for the Platte River near Ashland, Nebraska and Louisville, Nebraska, respectively. Gage 06801000 near Ashland, Nebraska was utilized primarily for site geometry and hydrology data. Water quality, riverbed characteristics, and river temperature data were obtained for gage 0680500 near Louisville, Nebraska, approximately 20 km downstream from the study location.

Russian River at Sonoma County, California. The primary source of information for the study site adjacent to the Russian River was the Sonoma County Water Agency (SCWA) and three USGS gages in Sonoma County. The three USGS gages utilized included gage 11464000 – Russian River near Healdsburg, California, gage 11465350 – Dry Creek Mouth near Healdsburg, California, and gage 11467000 – Russian River near Guerneville, California. Gage

1146700 – Russian River near Guerneville was utilized primarily for water quality data. Flows from gage 1145350 – Dry Creek Month near Healdsburg, California and gage 1146400 – Russian River near Healdsburg, California were summed to estimate the discharge at the wellfield. Cross-sections from the Sonoma County Aggregate Resources Management Plan (Sonoma County Permit and Resource Management Department and SCWA 2001) were used to estimate river geometry parameters. Extensive operational data for this site could not be obtained within the timeframe of the study. However, the average combined production from the two wellfields for periods when the diversion dam was lowered was available from Jasperse and Beach (2000).

Missouri River at Kansas City, Kansas. The main sources of information for the study site adjacent to the Missouri River at Kansas City, Kansas, included USGS, Kansas City Board of Public, Utilities, and Missouri Water Services Department. The USGS data were obtained for USGS gage 06893000 – Missouri River at Kansas City. Aquifer information was obtained from a USGS report (Kelly 2003) and from Collector Wells, INC (2005).

Study Location on the Llobregat, Rhine and Enns Rivers. Five of the 11 sites included in the study are located in Europe. Data from these sites was obtained primarily through email correspondence with experts at the site locations.

APPENDIX B
SHEAR STRESS ESTIMATION AT LOUISVILLE

The process of riverbed clogging can be offset by the regenerative process of streambed scouring. Schubert (2002a) has described the hydraulic process of streambed scouring and its importance in the limitation of clogging in RBF systems. Riverbed scouring is the result of shear forces imparted on a riverbed by the motion of the water passing over the riverbed, and the resistance to this motion imparted by the riverbed itself. While there is no practical technique for directly measuring this shear force on a riverbed, it can be estimated by the surface slope of the stream, vertical velocity profiles in the stream, and the sediment in transport on the riverbed. Each of these three techniques is examined in this appendix, using data from the Ohio River at Louisville, Kentucky as an example.

ESTIMATING RIVERBED SHEAR STRESS WITH STREAM SLOPE-LOUISVILLE

The shear stress acting on a riverbed manifests itself in several physically observable characteristics of a stream, the most obvious being the surface slope of the stream. The slope of the river surface undisturbed by falls or ripples provides a direct measure of the resistance to flow imparted by a streambed. In a perfectly uniform reach of a river, the stream slope provides a direct measure of riverbed shear forces. However, the presence of bends, bedforms (ripples and dunes on the streambed), and physical obstructions like bridge piers and driftwood also impart a resistance to flow, and the total stream energy loss inferred by stream slope may be as much as twice that exerted by streambed shear forces alone (Leopold, Wolman, and Miller 1992).

Extensive field data on stream surface elevations were collected on the Ohio River in the Louisville area by USGS for a stream modeling program funded by ORSANCO (Wagner and Mueller, 2001). This project included both field-measured surface elevations and flow-modeled surface elevations from which streambed characteristics may be computed. These data were supplemented with stream slope data gathered during this study from USGS, U.S. Army Corps of Engineers, National Oceanic and Atmospheric Administration, and Louisville Water Company sources for the April 2002 high flow condition experienced in the Ohio River.

These slope data were used to estimate the shear stress on the riverbottom according to the technique described in Julien (1998) as follows:

Shear stress = Slope * Density of water * depth

Available data from three stream flow conditions were evaluated for stream slope and computed shear stress. Between the points of measure in the river there are few bends and no bridges or dams, and thus the calculations were expected to give reasonable estimations of bed shear stress. These calculations are summarized in Table B.1.

The maximum estimated riverbed shear stress was 9.2 Newtons/m^2 (N/m^2). This compares to the average shear stress on the Rhine River of 10 N/m^2 as reported by Schubert (2002a).

Table B.1 Shear stress estimated from slope calculations

Location (river miles)	Flow m^3/sec	Slope 1/1000	Depth (meters)	Shear Stress (Newton/m^2)
High Flow Conditions				
589.8 to 594.6	10,847	0.043	9.6	4.05
594.6 to 599.5	10,847	0.051	12.3	6.13
599.5 to 604.8	10,847	0.047	13.4	6.16
560 to 600.6	10,818	0.058	11.1	6.31
600.6 to 604.8	10,818	0.036	13.9	4.90
560 to 600.6	13,282	0.084	11.1	9.16
Low Flow Conditions				
589.8 to 594.6	963	0.0031	7.6	0.24
594.6 to 599.5	963	0.0019	10.7	0.20
599.5 to 604.8	963	0.0093	12.2	1.11

ESTIMATING RIVERBED SHEAR STRESS FROM STREAM VELOCITY PROFILES

The shear forces are a function of stream velocity, and may be estimated by various stream velocity measures such as average stream velocity, which is often estimated by stream flow. The one-dimensional average stream velocity is a gross estimate of complex 3-dimensional in-stream velocities. Under ideal conditions, the velocity in a direction of flow perpendicular to the cross-section of the stream will demonstrate a profile versus depth that is logarithmic in shape with zero velocity at the riverbed (Julien, 1998).

The characteristics of this idealized profile provide an estimate of the shear stress imparted by the stream bed at that point. This technique of estimating shear stress on a streambed requires data on the velocity profile of the stream. Two hypothetical velocity profiles with zero velocity near the riverbed and a logarithmic increase in velocity with depth from the bottom are shown in Figure B.1, demonstrating relatively low shear stress from a smooth-bottom stream and higher shear stress from a rougher bottom stream.

A calculated shear velocity, shear stress, and friction slope can be derived from any two points on the velocity profile, according to the following equations (Julien 1998):

$$\text{Shear Velocity} = k^* \text{ Ln } (V_2\text{-}V_1)/\text{Ln } (D_2/D_1)$$

$$\text{Bed shear stress} = (\text{density of water})^*(\text{bed shear velocity})^2$$

$$\text{Friction slope} = (\text{shear stress})/(\text{spec wt water}^*\text{depth from bottom})$$

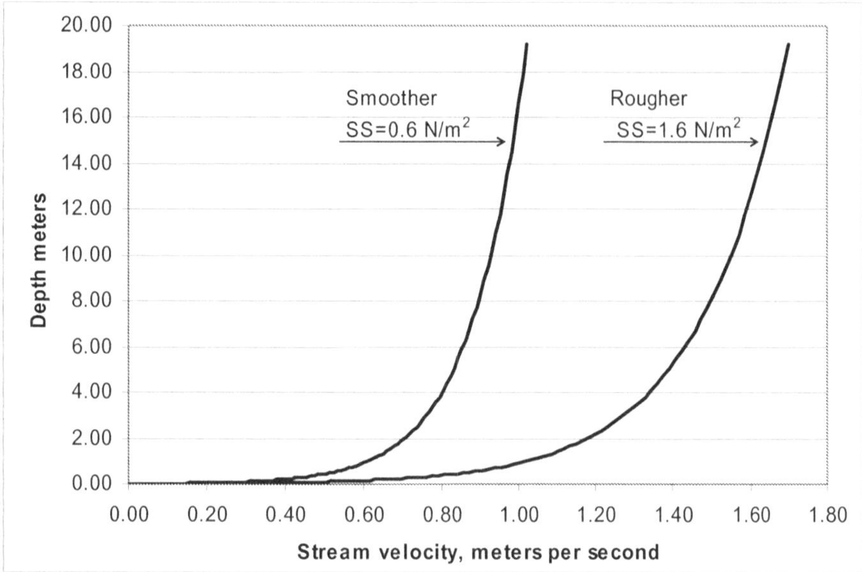

Figure B.1: Theoretical velocity profiles

In these equations, V_1 and V_2 are velocity at a given depths D_1 and D_2 from the stream bottom, *k* is the von karman constant of 0.4, the density of water is 1000 kg/m^3, and the specific weight of water is 9810 Newtons/ m^3 at 10 degrees Celsius.

Recent advances with in-stream velocity measurement with acoustic Doppler current profiler technology have provided huge databases of discrete velocity components across entire river cross-sections. These data allow the estimation of riverbed shear forces at specific locations, based on the velocity profile at that point. This technology suffers from the inability to produce data within 1 m of the stream bed and stream surface. The velocity readings near the stream bed are critical in estimating the riverbed shear stress, thus limiting the value of these data in estimating shear stress.

Velocity profiles were measured on the Ohio River in the Louisville area by USGS (Wagner and Mueller, 2001) using acoustic Doppler current profiler technology. Data selected to illustrate data manipulation techniques used in this appendix were collected at mid-stream at a river flow of 11,000 m^3/s and are presented in Table B.2. These data are instantaneous readings, and are thus subject to the variations caused by eddy currents. Ideally, time-averaged data would be used; however, the profiler data by its nature is instantaneous.

Because of the variability of the data and its departure from the theoretical log profile with zero velocity at the riverbed, some interpretation of the profile is required in order to estimate riverbed shear stress. The profile was fitted to a logarithmic curve using Excel's log estimate regression function, and values for shear velocity, shear stress, and friction slope were calculated. The profile was then modified with a zero velocity boundary and the log-fit curve data was again analyzed for shear velocity, stress, and slope. The various manipulations to the data are shown in Figure B.2, and a summary of these calculations for shear velocity, shear stress and friction slope is provided in Table B.3.

The most notable impact of these data manipulations was the assignment of a zero velocity near the stream bottom, which drove a much flatter slope as the velocity profile neared the y axis, compared to the raw data. This creates an assumed condition of a smoother boundary interface, compared to the shape of the profile without a forced zero intercept.

Table B.2 Velocity profile data from acoustic Doppler.
(Data from USGS)

depth	velocity
meters	m/sec
11.44	1.63
10.94	1.69
10.44	1.68
9.94	1.56
9.44	1.70
8.94	1.62
8.44	1.60
7.94	1.50
7.44	1.51
6.94	1.47
6.44	1.42
5.94	1.31
5.44	1.37
4.94	1.29
4.44	1.28
3.94	1.27
3.44	1.27
2.94	1.25
2.44	1.01
1.94	1.03
1.44	0.91
0.003	0.00
AVG	1.42

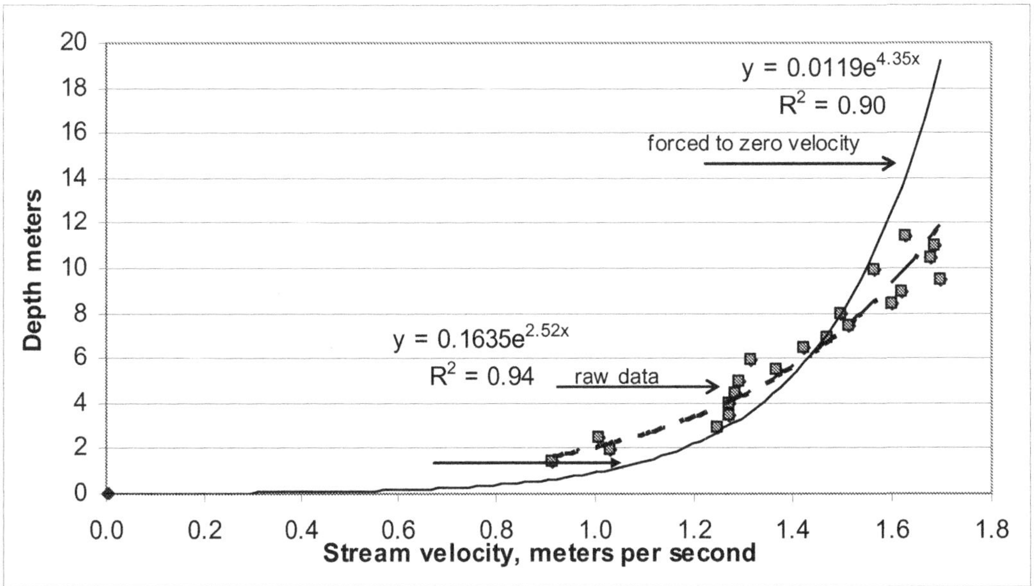

Figure B.2 Modifications of the velocity profile for shear stress calculations.

Table B.3 Shear stress calculation from modifications of velocity profile data

	Shear velocity	shear stress	friction slope
raw data log-fit	0.15	0.72	0.0064
raw data plus forced zero	0.08	0.21	0.0018
Bottom 20' plus zero	0.07	0.16	0.0014

Using the techniques described above, a complete cross-section of acoustic Doppler velocity profiles was analyzed. Data selected for evaluation were collected in the Ohio River adjacent to the RBF site at rivermile 595 at a river flow of 11,000 m^3/s (Wagner and Mueller, 2001). A total of 38 vertical profiles were evaluated from the river at this cross section. The number of velocity data points in each profile varied from 6 to 21; only two profiles (at each bank) contained fewer than 10 points. Most profiles contained at least 14 data points. The maximum shear stress calculated from the raw velocity profile data was 29 N/m^2; the minimum was 2.5 N/m^2; the average was 14 N/m^2, with a standard deviation of 5.8 N/m^2. These data are presented in the top graph of Figure B.3.

These profiles were modified by arbitrarily adding a near zero velocity (0.03 m/s) assigned to a depth of 0.03 m from the riverbed, as described above. These data are presented in the middle graph of Figure B.3. The impact of this modification to the velocity profile was significant, resulting in much less variation in estimated streambed shear stress across the river cross section. A comparison of the statistics from these two datasets is provided in Table B.4.

Data from three velocity profiles were selected for sensitivity analysis of the arbitrarily assigned depth and velocity values for the near-zero velocity datum. The selected velocity profiles included one each showing an increase, decrease, and no change in streambed shear stress as a result of the dataset manipulation with a near-zero datum of 0.03 m/s at 0.03 m above the streambed (middle figure dataset of Figure B.3). The selected velocity profiles are indicated in the bottom graph of Figure B.3, which shows the combined datasets for comparison. The data from the sensitivity analysis are provided in Table B.5. These data indicate that increasing the distance from the streambed assigned to a near-zero velocity (.003 m/s) from 0.003m to 0.03m resulted in an approximate doubling of the calculated shear stress. Increasing this distance from 0.03 m to 0.3 m from the streambed resulted in another approximate doubling of the calculated shear stress. At a depth of 0.03 m from the riverbed, decreasing the velocity from 0.03 to 0.003 m/s had no significant impact on the calculation of riverbed shear stress. The regression R^2 values comparing each set of Doppler data (raw and modified) to the theoretical values were highest for the near-zero value of 0.03 m/s at 0.03 m from the riverbed.

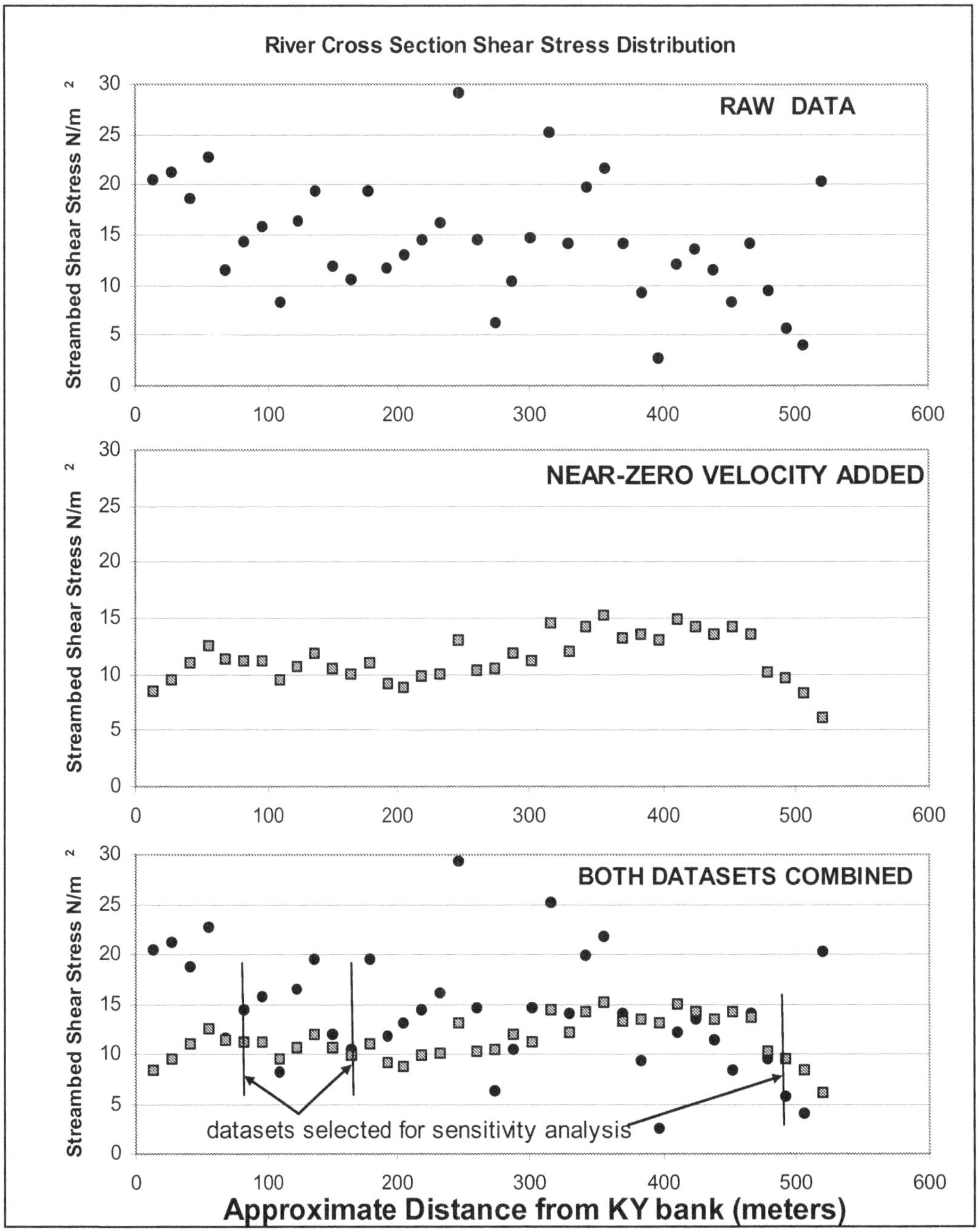

Figure B.3: Analysis of the impact of data manipulation on the estimation of streambed shear stress from velocity profiles.

Table B.4: Summary statistics for shear stress estimates derived from velocity profile datasets (all values are shear stress, expressed in N/m^2)

	Raw Data	Near-Zero Velocity Added
Average	14.3	11.2
Max	29.2	15.1
Min	2.5	6.0
Std Dev	5.9	2.2

Table B.5: Sensitivity analysis of the impact of varying added datum near zero.

SENSITIVITY ANALYSIS: Shear Stress (N/m^2)					
		Added datum (velocity m/s)/(depth m)			
DATASET	RAW	**.03/.03**	.003/.003	.003/.03	.003/.3
179	14	**11**	6	11	30
299	10	**10**	5	10	21
519	25	**14**	8	15	32
779	6	**10**	5	10	32

ESTIMATING RIVERBED SHEAR STRESS FROM STREAMBED SEDIMENT TRANSPORT

One of the more intuitive methods of assessing the shear stresses exerted on a streambed is to evaluate the size of the media that is in transport on the riverbed. While this can be accomplished by capturing media in transport during various flow events, it is also possible to get an indication of these shear stresses by simply observing the media on streambed itself under normal flow conditions.

Streambeds have naturally occurring and predictable variations in shear stress, depending on the stream geometry. It is easy to observe the impact and variations of shear forces on a stream by looking at a typical stream through a bend, pool, and riffle. Outside bends indicate high shear forces with steep cut walls and deposition of larger stones, or no deposition at all. Insides of bends show a gradation of smaller stones and gravels indicative of the lower velocities and lower shear stresses. Riffles reveal the larger stones that move only during high flow events, while bars show the impact of decreasing shear stresses downstream of bends and as the water levels recede after higher flow events. These patterns are consistent in both small and large streams.

It is also possible to use riverbed media to estimate shear forces resulting from recent flow events, as stones moved recently will be placed randomly on the streambed, thus exposing indications of having being recently moved. In areas where a particular size stone has recently moved, algae growth on both the top and bottom indicates recent movement (see Figure B.4. A stone with no algae present indicates that the stone moved from below the surface of the streambed, and was recently placed on the surface. A stone with algae on the bottom and not on the top indicates that the stone has not moved recently (see Figure B.5) By observing such deposition patterns in conjunction with recent streamflow events and measuring the stones on the riverbed, it is possible to estimate the relative shear stresses at various points in a stream, and with various flows.

An example of this shear stress estimating technique is provided for the Great Miami River in Cincinnati. Following a high flow event of 570 m^3/s, stones as large as 0.3 m by 0.1 m had rolled, as was indicated by algae on both the top and bottom of a stone as it lay in the stream. This stone was taken from a riffle and was typical of those found there. This circumstance would indicate the higher range of shear stresses for the stream for the given flow event. Most of the exposed bars were covered with medium cobble, roughly 0.1 m by 0.02 m, indicating

Figure B.4 Media on riverbed-note algae on large rock

Figure B.5 Rock not recently moved by shear forces-note algae on top, not on bottom

Figure B.6 Streambed rocks with orientation indicating anisotropy

Table B.6 Approximate threshold conditions for granular media at 20°C

Class name	d_s (mm)	Critical Shear Stress (N/m^2)
Large cobble	128	111
Small Cobble	64	53
Very Course Gravel	32	26
Course Gravel	16	12
Medium Gravel	8	5.7
Fine Gravel	4	2.7
Very Fine Gravel	2	1.3
Very Course Sand	1	0.47
Course Sand	.5	0.27
Medium Sand	.25	0.19
Fine Sand	.125	0.145
Very Find Sand	.0625	0.110

Data from Julien, 1998, *Erosion and Sedimentation*. Reprinted with the permission of Cambridge University Press

moderate or average shear forces for the streambed. The inside bends, conversely, had a wide range of stones on the surface, but only the smaller of these indicated recent movement (0.05 m and smaller).

Julien (1998) provides a compilation of data from the Highway Research Board, indicating the threshold shear forces required to move granular material in a streambed (Table B.6). These values indicate that the larger stones described above reflect shear stresses in the range of 200 N/m^2, while the cobble on the bars reflect shear stresses in the range of 80 N/m^2, and the gravel in the inside bends reflects shear stresses in the range of 40 N/m^2 .

This technique provides an opportunity to compare calculated shear stress estimates to estimates based upon riverbed composition, and supports the logical conclusion that the steeper streams with more dramatic flow variation will show higher and more variable shear stresses. This technique also provides information regarding the variation in streambed shear stress as a function of stream geometry within a given reach. This method of estimating riverbed shear stress is more easily applied to streams that provide easy access (streams with low flow depths of 1 m or less), although sediment samples from deeper streams may be similarly interpreted.

The picture in Figure B.6 also shows the natural orientation of riverbed media that produces anisotropic conditions in an aquifer. Random flipping of stones on the riverbed during high flow events results in random orientations with respect to resistance to streambed shear forces. Those stones in orientations exhibiting the greatest resistance to shear resist flipping, while stones in other orientations are more likely to flip again and again, until they are positioned in an orientation that resists flipping or are transported to an area of lower shear stress. This selectivity with respect to shear forces results in the pattern of consistent orientation and spacing of the stones on the riverbed as seen in the picture.

Referring back to the sediment data from the Ohio River at Louisville Figure 4.3, the bulk of the riverbed is composed of media with a D_{50} of 0.3 to 0.8 mm, indicating a shear stress at streambed transport of 0.2 to 0.4 N/m^2. The largest media on the streambed is in the range of 10 to 15 mm, indicating shear stresses of 6 to 10 N/m^2. These larger stones were found in a trough in the vicinity of a flow separation in the stream caused by upstream deposits from a small stream, and thus represent localized shear stresses.

The data from the Great Miami and Ohio Rivers illustrate how riverbed material analysis can be used to easily categorize the differences in streams with regards to shear stress. These differences are noticeable both within a given stream and between streams in general.

COMPARISON OF RIVERBED SHEAR STRESS ESTIMATION TECHNIQUES

The three techniques used to estimate the riverbed shear stresses (computation from stream surface slope, computation from velocity profiles, and inference from streambed media) provided similar estimates of the shear stresses acting on the riverbed, within reasonable accuracy. The computation from velocity profiles indicated a shear stress of approximately 11 N/m^2 as an average across the Ohio River near the LWC RBF facility. Stream surface slopes indicated shear stresses in the range of 4 to 6 N/m^2, and streambed particle sizes indicated that medium gravel (10 mm) moves on the riverbed, requiring shear stresses of at least 6 N/m^2.

Of these three techniques, the stream surface slope reflects not only riverbed flow resistance, but also any flow resistance imparted by bends, obstructions, drift materials, and any other impediment to flow. Because the estimate of streambed shear stress from stream surface slope assumes these values to be negligible, and attributes the slope and apparent energy dissipation only to bed shear stress, this estimate of bed shear stress should be higher than the estimates provided by the other techniques. The lack of significant bends in the river at this site and the relative consistency of the riverbed may explain the agreement in the results obtained from the three techniques used for estimating riverbed shear stress at this site.

The modification of the velocity profiles to include a near-zero velocity at the streambed surface greatly reduced the variation observed in the results of the shear stress computations based on velocity profiles. The velocity profiles were altered twice in this analysis: initially by fitting an exponential equation to each profile, and then by adding the near-zero velocity and re-fitting the exponential equation. The first modification tended to distribute the error inherent in the discrete velocity readings obtained by means of acoustic Doppler technology. The addition of a near-zero velocity near the streambed is supported by fluid flow theory, although the value assigned (0.03 m/s at a point 0.03 m from the riverbed) was arbitrary. The regression fit coefficient R^2 value for each profile was consistently higher for this modified dataset, typically in the mid-to-upper 90 percent range, than for the raw dataset; the R^2 value was lower and more variable in each dataset when the other near-zero values of velocity were assumed for the riverbed (Table B.5). It thus appears that there may be justification in modifying the acoustic Doppler dataset for a near-zero streambed velocity value, as the data better fit the theoretical flow regime, the estimate was closer to the values obtained by stream slope calculations, and the values more closely fit the shear stress implied by streambed particle size. Ideally, however, a dataset that includes several datapoints nearer the riverbed would be used for these calculations.

APPENDIX C
DATA FROM SEEPAGE METER

Ohio River Infiltration Rate Testing Party: MDU, DDZ, DLM, MDK, Steve Hubbs, Tiffany Caldwell

Date: August 16, 2004

River conditions: Velocities estimated at 0.4 mph (NOAA); stage near 420 ft msl

Location		Depth of	Time of	Test	Run	Initial	Remaining	Infiltration	Run	Rate	Rate	Hose	Remarks
Latitude	Longitude	water	day	#	#	volume	volume	volume	time			length	
		(ft)				(mL)	(mL)	(mL)	(min)	(mL/min)	ft/hour	(ft)	
38 21 2.3	085 38 14.5	30	1320	1	1	4000	3840	160	3	53	0.11	43	
			1330	1	2	4000	3230	770	10	77	0.16	43	
			1345	1	3	4000	4070	-70	10	-7	-0.01	43	
			1400	1	4	4000	4120	-120	5	-24	-0.05	43	
			1410	2	1	*4000*	440	*3560*	5	712	1.51	43	Reset can at same location
			1415	2	2	4000	2570	1430	65 sec	**1320**	**2.80**	43	420 ft off KY bank
			1420	2	3	4000	2160	1840	1.5	**1230**	**2.61**	43	
			1430	2	4	4000	1500	2500	2	**1250**	**2.65**	43	**1266 2.68 average of BOLD****
			1440	2	5	4000	1470	2530	2	**1265**	**2.68**	43	**ml/min ft/hour 0.818 m/hr**

38 21.025	085 38.230	29	1500	3	1	4000	5100	-1100	3	-367	-0.78	43	
			1510	4	1	4000	5480	-1480	3	-493	-1.04	43	Reset can at same location
			1520	5	1	4000	4840	-840	3	-280	-0.59	43	Reset can at same location

** For this test, the best values to use are those with a time under 2 minutes, as the reset gave highly reproducible results

Ohio River Infiltration Rate Testing   Party: MDU, DDZ, DLM, MDK, Steve Hubbs

Date: August 17, 2004

River conditions: Velocities estimated at 0.5 mph (NOAA); stage near 420 ft msl

Location Latitude	Longitude	Depth of water (ft)	Time of day	Test #	Run #	Initial volume (mL)	Remaining volume (mL)	Infiltration volume (mL)	Run time (min)	Rate (mL/min)	Rate ft/hour	Hose length (ft)	Remarks
38 21 0.94	085 38 13.87	30	930	1	1	4000	4480	-480	5	-96	-0.20	43	
			935	2	1	4000	4000	0	5	0	0.00	43	Reset can at same location
			945	2	2	4000	4000	0	5	0	0.00	43	
38 21 1.26	085 38 14.68	30	1015	3	1	*4000*	3380	*620*	2.5	248	0.53	43	
			1020	3	2	4000	3280	720	5.5	131	0.28	43	
			1035	3	3	4000	3700	300	0.382	32.7	0.07	43	
				3	4	4000	4050	-50	10	-5	-0.01	43	Reset can at same location
			1100	3	5	4000	3960	40	10	4	0.01	43	
38 21 1.44	085 38 15.30	31	1140	4	1	4000	770	3230	4	**808**	**1.71**	43	435 ft from KY bank
			1145	4	2	4000	1420	2580	3	**860**	**1.82**	43	
			1150	4	3	4000	2270	1730	2	**865**	**1.83**	43	**848.25 1.797 average of BOLD**
			1200	4	4	4000	2280	1720	2	**860**	**1.82**	43	**ml/min ft/hour 0.54773 m/hr**
38 21 1.92	085 38 15.62	32	1250	5	1	4000	3480	520	1.5	**346.7**	**0.73**	43	161 yds. Off KY 483 feet
			1300	5	2	4000	2320	1640	5	**328**	**0.69**	43	
			1310	5	3	4000	2180	1820	5	**364**	**0.77**	43	**345.68 0.732 average of BOLD**
			1315	5	4	4000	2280	1720	5	**344**	**0.73**	43	**ml/min ft/hour 0.223208 m/hr**
38 21 2.87	085 38 16.44	34	1345	6	1	4000	4960	-960	10	-96	-0.20	43	
			1400	6	2	4000	3900	100	10	10	0.02	43	
			1415	6	3	4000	3800	200	10	20	0.04	43	
			1430	6	4	4000	3980	20	10	2	0.00	43	
			1445	6	5	4000	4020	-20	10	-2	0.00	43	

Ohio River Infiltration Rate Testing　　Party: MDU, DDZ, DLM, MDK, Steve Hubbs

Date: August 18, 2004

River conditions: Velocities estimated at 0.6 mph (NOAA); stage near 420 ft msl

Location Latitude	Longitude	Depth of water (ft)	Time of day	Test #	Run #	Initial volume (mL)	Remaining volume (mL)	Infiltration volume (mL)	Run time (min)	Rate (mL/min)	Rate ft/hour	Hose length (ft)	Remarks
38 21 1.46	085 38 14.15	29	950	1	1	4000	6000	-2000	0.257	-324	-0.69	43	118 yds. Off KY bank
			1005	2	1	4000	1620	2380	5	**476**	**1.01**	43	Reset can at same location
			1015	2	2	4000	2540	1460	5	**292**	**0.62**	43	Windy and choppy
			1020	2	3	4000	2360	1640	5	**328**	**0.69**	43	Severe wind and chop
			1030	2	4	*4000*	3150	*850*	5	170	0.36	43	White caps on river
			1035	2	5	4000	2360	1640	**5**	**328**	**0.69**	43	**356 ml/min 0.75 ft/hour Average of BOLD**
			1045	2	6	4000	5400	-1400	**5**	-280	-0.59	43	**0.230 m/hr**
38 21 1.85	085 38 14.22	29	1150	3	1	4000	2770	1230	5	**246**	**0.52**	43	Boats pointing downstream:
			1200	3	2	4000	3200	800	5	**160**	**0.34**	43	GPS 20' to KY side
			1210	3	3	4000	2740	1260	5	**252**	**0.53**	43	**219 ml/min 0.46 ft/hour Average of BOLD**
			1215	3	4	4000	3820	180	5	36	0.08	43	**0.142 m/hr**
			1250	3	5	4000	5030	-1030	5	-206	-0.44	43	
			1305	4	1	4000	4300	-300	5	-60	-0.13	43	Reset can at same location

Continued field work with penetrometer readings

Ohio River Infiltration Rate Testing   Party: MDU, DDZ, DLM, MDK

Date: August 19, 2004

River conditions: Velocities estimated at 0.6 mph (NOAA); stage near 420 ft msl

Location Latitude	Longitude	Depth of water (ft)	Time of day	Test #	Run #	Initial volume (mL)	Remaining volume (mL)	Infiltration volume (mL)	Run time (min)	Rate (mL/min)	Rate ft/hour	Hose length (ft)	Remarks
38 21 7.51	085 38 12.28	28	945	1	1	4000	4000	0	5	0	0.00	43	138 yds. Off the KY bank
			955	1	2	4000	3690	310	5	**62**	**0.13**	43	
			1005	1	3	4000	3760	240	5	**48**	**0.10**	43	**50 0.11 average of BOLD**
			1010	1	4	4000	3800	200	5	**40**	**0.08**	43	**ml/min ft/hour 0.032 m/hr**
38 21 4.76	085 38 13.36	33	1040	2	1	4000	3140	860	**5**	**172**	**0.36**	43	141 yds. Off KY bank
			1050	2	2	4000	3080	920	**5**	**184**	**0.39**	43	**180 0.38 average of BOLD**
			1100	2	3	4000	3080	920	**5**	**184**	**0.39**	43	**ml/min ft/hour 0.116 m/hr**
38 21 0.12	085 38 16.58	32	1135	3	1	4000	3800	200	5	40	0.08	43	
			1145	3	2	4000	3720	280	5	56	0.12	43	
			1150	3	3	4000	3370	630	5	126	0.27	43	
			1200	3	4	4000	6000	-2000	5	-400	-0.85	43	**Can dragged on riverbottom**
			1210	4	1	4000	3760	240	5	48	0.10	43	Reset can at same location
38 20 56.43	085 38 17.60	34	1315	5	1	4000	4000	0	5	0	0.00	43	
			1325	5	2	4000	4000	0	5	0	0.00	43	
			1335	5	3	4000	3900	100	5	20	0.04	43	**Reset can at same location**
			1345	5	4	4000	3820	180	5	36	0.08	43	
			1400	5	5	4000	4000	0	5	0	0.00	43	
38 21 4.59	085 38 10.85	20	1430	6	1	4000	4000	0	5	0	0.00	43	
			1440	6	2	4000	4000	0	5	0	0.00	43	
			1445	6	3	4000	4000	0	5	0	0.00	43	
38 21 2.22	085 38 11.71	28	1530	7	1	4000	3400	600	5	120	0.25	43	
			1535	7	2	4000	3600	400	5	80	0.17	43	
			1540	7	3	4000	1600	2400	5	480	1.02	43	
			1545	7	4	4000	3200	800	5	160	0.34	43	
			1555	7	5	4000	2300	1700	5	340	0.72	43	
			1600	7	6	4000	4000	0	5	0	0.00	43	
			1610	7	7	4000	5400	-1400	5	-280	-0.59	43	

Ohio River Infiltration Rate Testing   Party: MDU, DDZ, DLM, MDK, Steve Hubbs

Date: August 20, 2004

River conditions: Velocities estimated at 0.9 mph (NOAA); stage near 420 ft msl

Location Latitude	Longitude	Depth of water (ft)	Time of day	Test #	Run #	Initial volume (mL)	Remaining volume (mL)	Infiltration volume (mL)	Run time (min)	Rate (mL/min)	Rate ft/hour	Hose length (ft)	Remarks
38 20 58.42	085 38 14.15	33	940	1	1	4000	4570	-570	5	-114	-0.24	43	
			950	2	1	4000	4000	0	5	0	0.00	43	Reset can at same location
			1000	2	2	4000	5360	-1360	5	-272	-0.58	43	
			1010	3	1	4000	4000	0	5	0	0.00	43	Reset can at same location
			1015	3	2	*4000*	4000	*0*	5	0	0.00	43	Penetrometer reading - 4 to 6 in.

38 21 0.00	085 38 13.44	33	1100	4	1	4000	4000	0	5	0	0.00	43	
			1115	4	2	4000	4600	-600	5	-30	-0.06	43	
			1120	5	1	4000	4600	-600	5	-30	-0.06	43	Reset can at same location

Suspended field operations due to rain and worsening river conditions

APPENDIX D
EVALUATION OF 1999 AND 2004 PUMP TESTS

Pump tests were conducted in June 1999 during the initial start-up of the RBF system, and July 2004 after the system had been idle for a period of 4 months. High-frequency data collection extended for several weeks during these tests. Data were used to evaluate piezometric surfaces and estimate hydraulic conductivity of the riverbed.

START-UP PUMP TEST, JUNE 1999

Data from the initial start-up was used to estimate initial travel velocities across the riverbed and through the aquifer as a part of an investigation into water quality changes occurring in the RBF process (Wang, Hubbs, and Song 2002). These data were re-analyzed in this study to provide the vertical velocity component and headloss across the distance traveled to estimate riverbed hydraulic conductivity in the riverbed prior to the influence of clogging. Data from the first 60 hours of this test are presented in Figure D.1. The velocity was estimated as the time required to reach the midpoint of the temperature breakthrough curve.

START-UP PUMP TEST, JULY 2004

The collector well was shut down from April through July 28, 2004 for maintenance. This four month period of idle allowed a temperature difference between the aquifer and river to re-establish, and the pump test of June 1999 was repeated. The results, however, were strikingly different, as is illustrated in Figure D.2. The temperature breakthrough curve at 2.7 m below the riverbed indicated that the conditions controlling the velocity of water in 1999 had changed significantly in 2004. Similar to the pump test in 1999, the majority of the head loss occurred in the first 0.6 m of the riverbed. The head loss after 48 hours of pumping was approximately 3 m greater in 2004 that in 1999.

Figure D.3 shows the temperature and head data for the 2004 test over a 3 week period, to the point where the temperature data stabilized with the river temperature. Note that the riverbed elevation at the point where the probes are located was approximately 122.5 m amsl when the probes were set in 1999. The streambed has been modified significantly by sediment erosion and deposition over the 5 year period. The general pattern is erosion in the spring, and deposition through the remainder of the year. In June, 2004, the approximate elevation near the location of the probes was 122 m amsl.

Interpreting the temperature and pressure head data for 2004 in relation to the depth of the riverbed at the point of measurement, a dramatic change in velocity occurred near to the time when the pressure head measured at the probes approached the elevation of the riverbed over the probes. The pressure head measured at the probes was at elevation 123 m amsl (very near the riverbed elevation) approximately 10 hours after the start of the 2004 pump test at which time dramatic changes in the temperature breakthrough curves were observed in Figure D.3. These data are strongly suggestive that the change in riverbed conductance was tied directly to the passage of the piezometric surface under the riverbed.

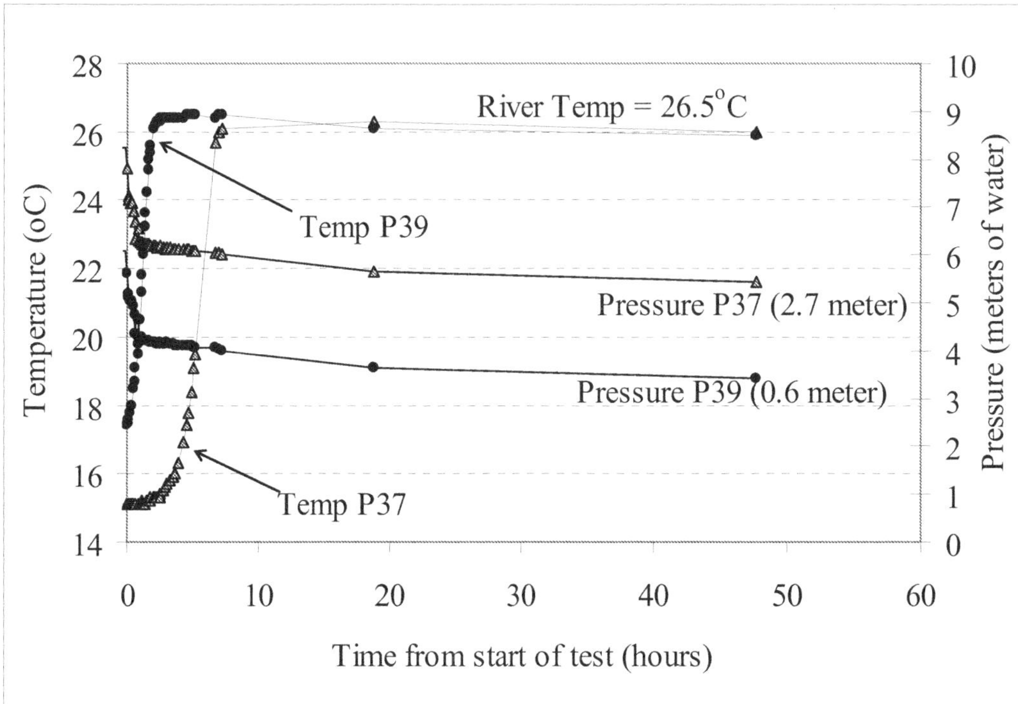

Figure D.1 Pressure and temperature versus time, initial pump start-up June 1999

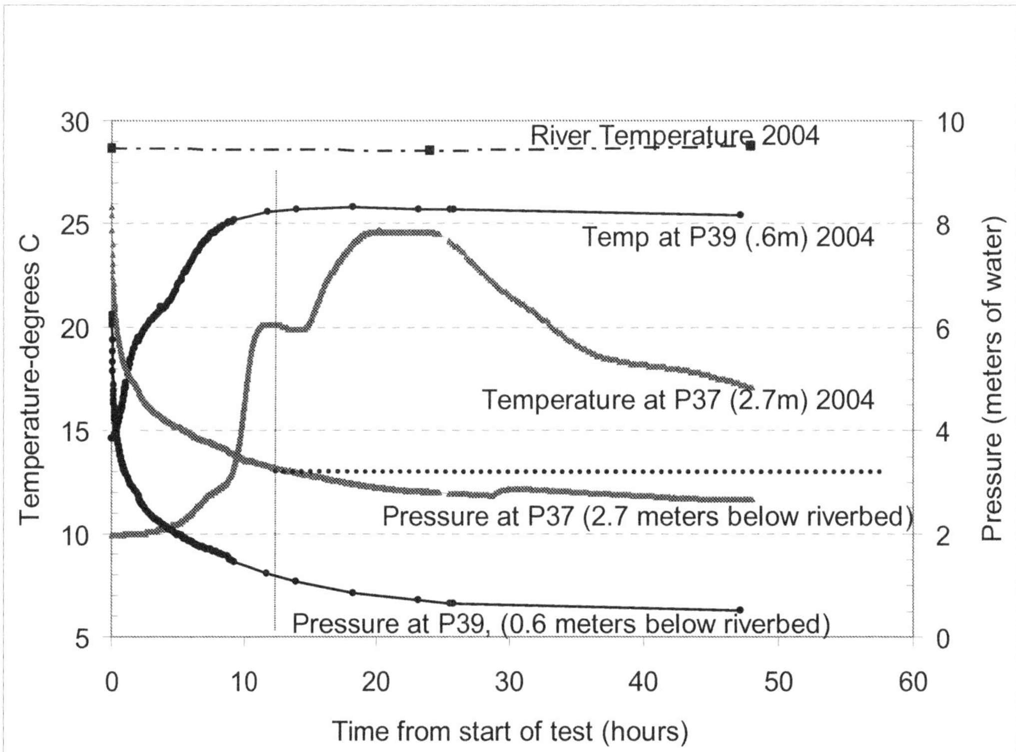

Figure D.2 Pressure and temperature versus time, July 2004 pump test.

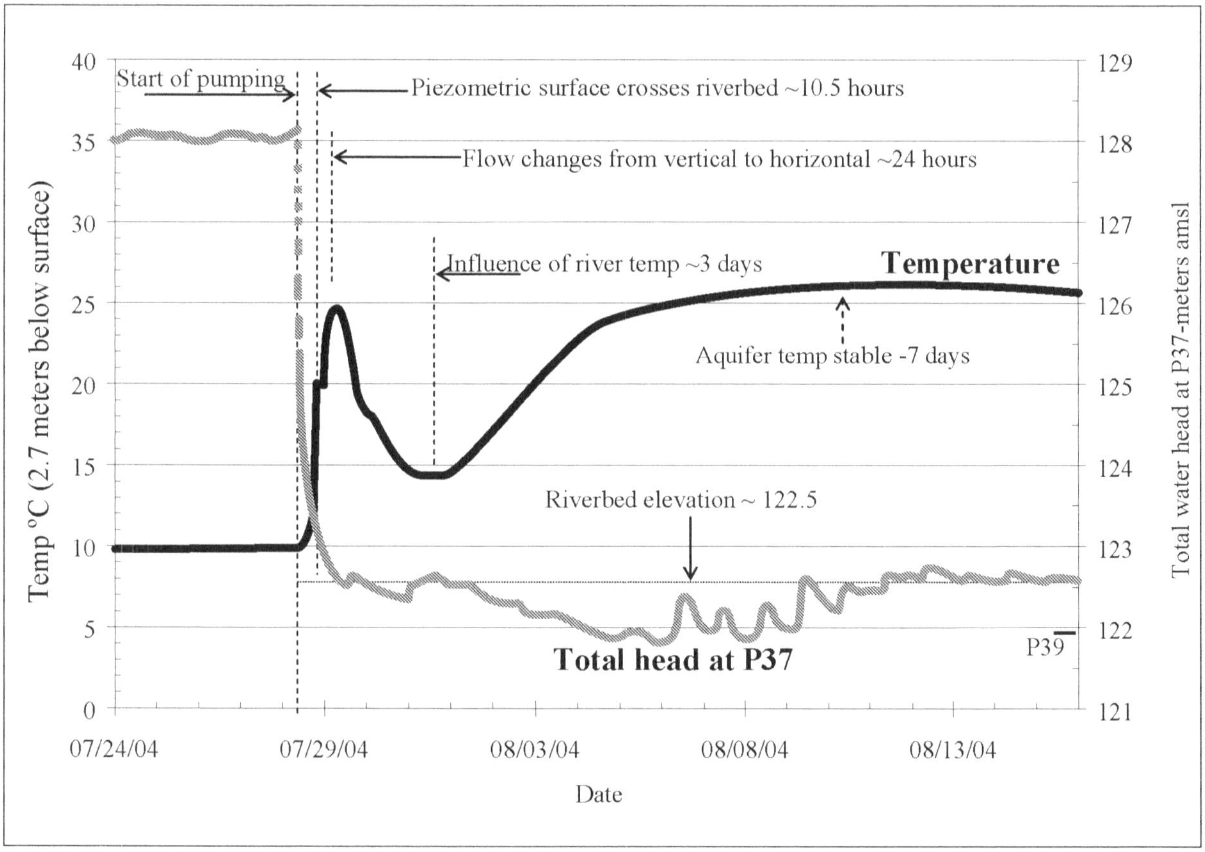

Figure D.3 Pressure and head measurements during the first 4 weeks of pumping of the July 2004 pump test.

INTERPRETING THE 2004 PUMP TEST: THE INFLUENCE OF UNSATURATED CONDITIONS IN THE AQUIFER

If it is assumed that changes in the temperature breakthrough curve in shown in Figure D.3 are the result of changes in riverbed hydraulic conductivity, then the shape of the curve can be explained as follows:

- From the start of the test through hour 10, the flow through the riverbed was primarily vertical, with the temperature reflecting water velocity as the river water penetrated the aquifer.
- The colder aquifer water temperature and riverbed clogging resulted in greater drawdown in the well and lower pressures in piezometers P37 and P39 compared to 1999 data.
- Between hours 10 and 11, the piezometric surface as measured at P37 and P39 dropped below the elevation of the riverbed, resulting in increased pressure on the riverbed as unsaturated conditions began to develop. The resulting decrease in hydraulic conductivity greatly decreased the vertical flow into the riverbed above the probes.
- Between hours 10 and 15, flow continued past the lower probe P37 in a vertical direction, as the aquifer above the probes drained towards the well. Very little flow was entering the aquifer immediately above the probes.
- Between hours 15 and 24 the deeper probe (probe P37 2.7 m below the riverbed) was influenced by flow with an increasing horizontal component, as the riverbed impacted by decreased hydraulic conductivity (clogged zone) expanded.

- Between hours 24 and 72, the flow past the probe was primarily horizontal, with colder aquifer water from further away from the well passing under the clogged zone, dropping the temperature at probe P37.
- Between days 3 and 7, the temperature at probe P37 reflected the impact of river recharge from a point distant from P37. Flow was primarily horizontal, with water temperature increasing with time reflecting the impact of river temperature at the point of recharge. The temperature at P37 reflected river temperature with a lag time of approximately 7 days.

This scenario is possible and reasonable, given the observations from this pumping test and indications of changes in riverbed flux rates observed as a function of distance from the well. A series of aquifer temperature profiles was developed to demonstrate the flow of water and changes in temperature that might have been expected under the above scenario. Only the temperature data at the two probes represent actual data. The shapes attributed to the temperature profiles are based upon anticipated flowpaths. These temperature profiles are presented in Figure D.4 and Figure D.5.

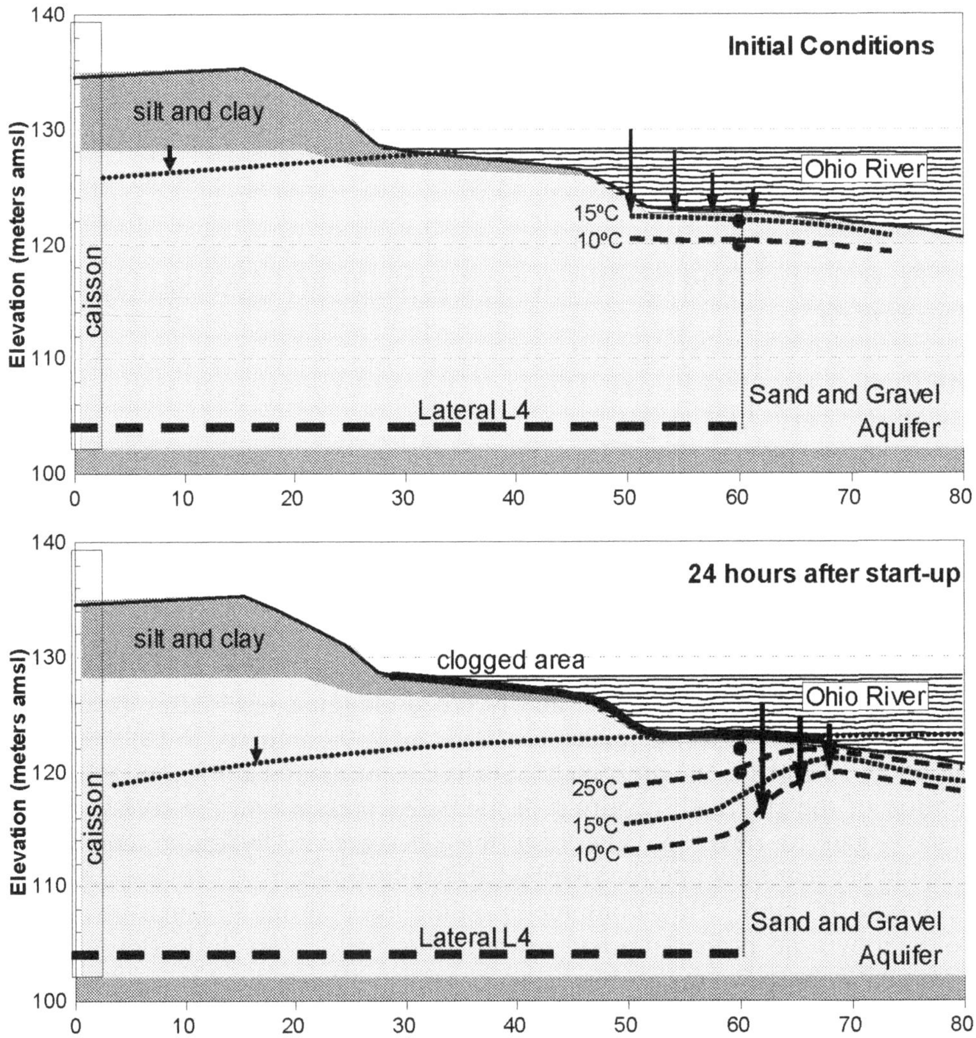

Figure D.4: Hypothetical temperature profiles for July 2004 start-up (cont next page).

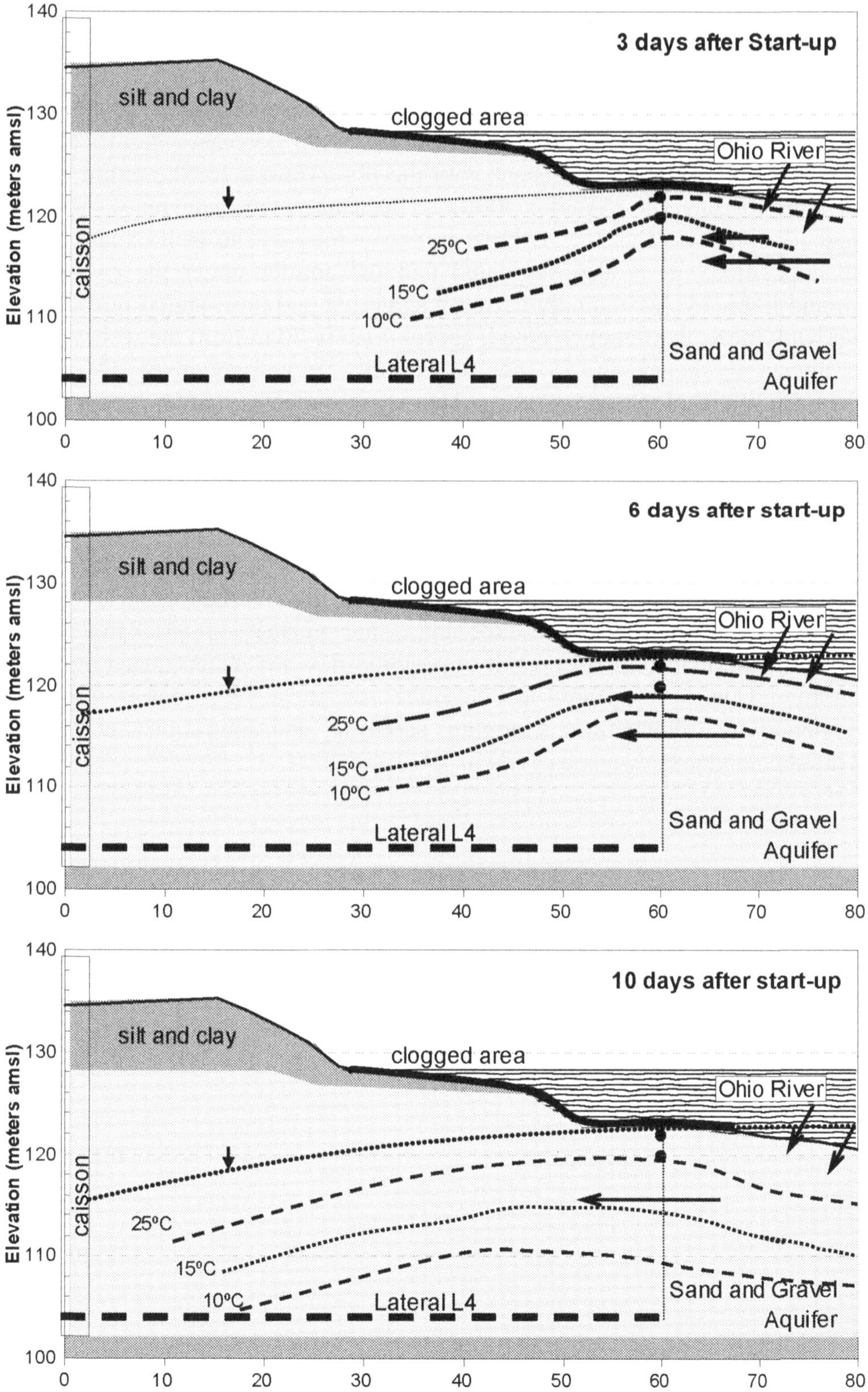

Figure D.5: Hypothetical temperature profiles for July 2004 start-up: After 6 days and after 10 days (continued from previous page).

CHANGES IN RIVERBED HYDRAULIC CONDUCTIVITY: 1999 AND 2004

The comparison of data from the 1999 and 2004 datasets leads to the conclusion that the riverbed hydraulic conductivity at the point of observation decreased significantly over the 5 year period. This conclusion is consistent with the observations for the entire riverbed system made by Schafer for this site (Schafer 2006). The cause of this decrease, however, remains undetermined. It is known that the riverbed in the vicinity of the well experiences significant erosion and deposition. It is also suspected that the aquifer under the riverbed is at times unsaturated, and likely cycles annually between saturated and unsaturated in the area where the sloping riverbank meets the flatter riverbed, approximately 70 m from the well.

The impact of the change in specific capacity as a function of distance to point of recharge has been modeled (Schafer 2006). This work indicates that an increase in distance of 60 m to the point of recharge would be expected to decrease well capacity by approximately 10%. Thus, the progression of a zone of decreased conductivity from 60 m to 120 m into the riverbed could account for approximately one third of the total reduction in specific capacity observed at this site.

The highly variable condition of the riverbank/riverbed interface at this site further complicates the interpretation of data. That portion of the riverbed experiencing the greatest changes with regards to saturation also experiences the greatest annual changes in bathemetry, erosion, and sediment deposition as a result of periodic flow reversals caused by irregularities in the riverbottom . Thus, it is difficult to determine how the several factors impacting riverbed hydraulic conductivity (temperature, clogging, scouring, and saturation) interacted to impact riverbed hydraulic conductivity as observed in 1999 and 2004.

APPENDIX E
TABLE OF CONVERSIONS

	To Convert		
Paramter	**From SI**	**To English**	**Multiply By**
Length	**meter**	**feet**	**3.281**
	cm	**inch**	**0.3937**
Flow	**m^3/day**	**mgd**	**.0002641**
	m^3/sec	**ft^3/sec**	**35.31**
Pressure	**N/m^2 (pascal)**	**psi**	**0.000145**
Specific Capacity	**m^2/d**	**mgd/ft**	**0.0000805**

REFERENCES

Baveye, P., P. Vandevivere, B.Hoyle, P. DeLeon, and D. Sanchez de Losada, 1998. Environmental Impact and Mechanisms of the Biologcial Clogging of Satuated Soils and Aquifer Materials, *Critical Reviews in Environmental Science and Technology*, 28 (2);123-191.

Collector Wells International, INC. 2005. Services-Specialty Expertise. [Online]. Available: <http://www.collectorwellsint.com/services.htm>. [cited March 1 2005].

Constantz, J. G. Su, and C. Hatch, 2006. Use of Heat as a Tracer for Examining Streambed Conductance at the Russian River RBF Site, Sonoma County, California. In *Riverbank Filtration Hydrology*. Edited by S. Hubbs. Dordrecht, The Netherlands: Springer.

Davis, M., and D. Cornwell, 1998. *Introduction to Environmental Engineering*, New York: McGraw Hill.

Domenico, P. and F. Schwartz, 1998. Physical and Chemical Hyrdogeology, New York: John Wiley and Sons, Inc.

Engesgaard, P. 2006. Bioclogging of Porous Media: Tracer Studies. In *Riverbank Filtration Hydrology*. Edited by S. Hubbs. Dordrecht, The Netherlands: Springer.

Freeze, R. and J. Cherry, 1979. Groundwater, Upper Saddle River, NJ: Prentice Hall Inc.

Gollnitz, W., B. Whitteberry, and J. Vogt, 2004. Riverbank Filtration: Induced Infiltration and Groundwater Quality. *Journal AWWA*. 96(12): 98-110.

Gollnitz, W., B. 2004. Personal Communication. Cincinnati Water Works.

Gollnitz, W., J. Clancy, B. Whitteberry, and J. Vogt, 2003. RBF as a Microbial Treatment Process. *Journal AWWA*. 95(12): 56-66.

Henson, J., J. Obrist, and K. Prior, K. [N.d.] Two Wells Yield 35 mgd: Lincoln's Wellfield Expansion.

Hubbs, S. 2006. Processes Involed in Mechanical Clogging of Riverbank Filtration Systems. In *Riverbank Filtration Hydrology*. Edited by S. Hubbs. Dordrecht, The Netherlands: Springer.

Hubbs, S. 2004. High Yield Riverbank Filtration Systems for Public Water Supply: Measuring the Impact of Riverbed Plugging. In Proceedings *of the 2004 World Water and Environmental Resources Congress. Critical Transitions in Water and Environmental Resources Management.* Reston, VA: ASCE. pp. 1-8.

Hubbs, S. 2002a. RBF Construction Options Considered at Louisville, KY. In *Riverbank Filtration –Improving Source Water Quality*, National Water Research Institute, Dordrecht, The Netherlands: Kluwer Academic Publishing.

Hubbs, S. 2002b. Defining Sediment Transport in the Ohio River at Louisville in the Context of Impact of Yield from Bank Filtration Systems. In-house report, Louisville Water Company, Louisville KY.

Jasperse, J., and R. Beach, 2000. An Analysis of the Water Production Capacity of the Sonoma County Water Agency Diversion Facilities without the Diversion Dam. Prepared in Cooperation with the Sonoma County Water Agency.

Julien, P. 1998. *Erosion and Sedimentation*. Cambridge, England: Cambridge University Press.

Kelly, B. 2003. Missouri River Alluvial Aquifer Ground-Water Protection. [Online]. Available: <http://mo.water.usgs.gov/indep/kelly/mo-alluvial-gw> [cited March 31, 2005].

Knighton, D. 1998. *Fluvial Form & Processes*. New York, NY: Oxford University Press, Inc.

Lee, E. <Elee@ci.lincon.ne.us>. 2004. Data for RBF Project. Email to Tiffany Caldwell <tiffany.caldwell@louisville.edu>.

Leopold, L.B., M. Wolman, and J. Miller, 1992. *Fluvial Processes in Geomorphology.* Mineola, New York: Dover Publications, Inc.

Martín-Alonso, J. 2006. Managing Resources in a European Semi-Arid Environment: Combined Use Of Surface and Groundwater for Drinking Water Production in the Barcelona Metropolitan Area. In *Riverbank Filtration Hydrology*. Edited by S. Hubbs. Dordrecht, The Netherlands: Springer.

Martín-Alonso, J. < jma@agbar.es>, 2005. Re: Riverbed Conductivity Data. E-mail to Tiffany Caldwell <tiffany.caldwell@louisville.edu>.

Martín-Alonso, J. < jma@agbar.es>, 2004. Re: Data for RBF sites in Spain. E-mail to Tiffany Caldwell <tiffany.caldwell@louisville.edu>.

Martin, M. <Mmartin@ci.lincoln.ne.us>, 2005. Additional data request. E-mail to Tiffany Caldwell <tiffany.caldwell@louisville.edu>.

Maynord, S. T., 1990. Velocities Induced by Commercial Navigation. Technical Report HL-90-15. Vicksburg, MS: U.S. Army Engineer Research Waterways Experiment Station.

Maynord, S. T., 2000. Shear Stress on Shallow-Draft Barge Hulls. ENV Report 24. Vicksburg, MS: Army Engineer Research Waterways Experiment Station.

National Water Research Institute, 2005. The Third International Riverbank Filtration Conference. [Online]. Available: <http://216.133.236.158.asp.sp.asp?main=m3&sub=s4&id=3>. [cited January 15, 2005].

National Weather Service (NWS) 2004. River Forecast Center. [Online]. Available: <http://www.crh.noaa.gov/ahps/rfc.php> [accessed October 22, 2004].

Schafer, D. 2000. Hydraulics Analysis of Groundwater Extraction at the B.E.Payne Water Treatment Plant. Louisville Water Company Report, Louisville KY, July 2000.

Schafer, D. 2006. Use of Aquifer Testing and Groundwater Modeling to Evaluate Aquifer/River Hydraulics at Louisville Water Company, Louisville, Kentucky, U.S.A.. In *Riverbank Filtration Hydrology*. Edited by S. Hubbs. Dordrecht, The Netherlands: Springer.

Schubert, J. 2006. Significance of Hydrologic Aspects on RBF Performance. In *Riverbank Filtration Hydrology*. Edited by S. Hubbs. Dordrecht, The Netherlands: Springer.

Schubert, J. <juschubert@mac.com>. 2004. Re: Data for workshop chapter. E-mail to Tiffany Caldwell <tiffany.caldwell@louisville.edu>.

Schubert, J. 2002a. Hydraulic aspects of riverbank filtration – field studies. *Journal of Hydrology.* 266: 145-161.

Shubert, J. 2002b. German Experience with Riverbank Filtration Systems. In *Riverbank Filtration –Improving Source Water Quality.* National Water Reseacrh Institute. Dordrecht, The Netherlands: Kluwer Academic Publishing.

Sheets, R., R. Darner, and B. Whitteberry, 2002. Lag times of bank filtration at a well field, Cincinnati, Ohio, U.S.A.. 1998. *Journal of Hydrology.* 266: 162-174.

Sonoma County Permit and Resource Management Department and Sonoma County Water Agency, 2001. Sonoma County Aggregate Resources Management Plan 1997 and 1998 Monitoring Program Results: Russian River, Sonoma County, California.

Stuyfzand, P., M. Juhàsz-Holterman, and W. de Lange, 2006. Riverbank Filtration in the Netherlands: Well Fields, Clogging and Geochemical Reactions. In *Riverbank Filtration Hydrology*. Edited by S. Hubbs. Dordrecht, The Netherlands: Springer.

USGS. 2004. NWIS Web Data for a Nation. [Online]. Available: <http://waterdata.usgs.gov/nwis> [cited October 22, 2004].

Vandevivere, P., and P. Baveye, 1992. Effect of Bacterial Extracellular Polymers on the Saturated Hydraulic Conductivity of Sand Columns. *Applied and Env. Microbiology*, May 1992: 1690-1698.

Wagner, C., and D. Mueller, 2001. Calibration and Validation of a Two-Dimensional Hydrodynamic Model of the Ohio River, Jefferson Co. KY. Water Resources Investigations Report 01-4091. Louisville, KY.: U.S. Dept of the Interior USGS.

Wang, J., S. Hubbs, and R. Song, 2002. Evaluation of Riverbank Filtration as a Drinking Water Treatment Process. Denver, CO.: AwwaRF and AWWA.

Wett, B. 2004. Monitoring clogging after start-up of a RBF-system at the River Enns, Austria. In *Riverbank Filtration Hydrolog*y. Edited by S. Hubbs. Dordrecht, The Netherlands: Springer.

WorldAtlas.Com. 2005). Major Rivers: Europe. [Online]. Available: <http://www.worldatlas.com/webimage/country/euriv.htm>. [cited March 1, 2005].

Yongstrom, G. <gregy@orsanco.org.>. 2005. Water Quality Data for Ohio River at Louisville. Email to <tiffany.caldwell@gmail.com>.

ABBREVIATIONS

amsl	above mean sea level
AWWA	American Water Works Association
AwwaRF	Awwa Research Foundation
^oC	degrees Celcius
C_s	specific capacity
D	diameter
deg	degree
DMWW	Des Moines Water Works
DOC	dissolved organic carbon
eds	Editors
^oF	degrees Fehrenheit
ft	foot or feet
ft^2	square foot or square feet
ft^3	cubic foot or cubic feet
ft^3/hr	cubic feet per hour
ft^3/min	cubic feet per minute
gal/ft/d	gallon per foot per day
γ	specific weight of water
gph	gallon per hour
gpm	gallon per minute
gpm/ft^2	gallon per minute per square foot
h	hour
hr	hour
in^2	square inch
in	inch
km	kilometer
K_{rb}	riverbed hydraulic conductivity
L	Liters
Ln	natural log
LWC	Louisville Water Company
m	meter
m^2	square meter

m^3	cubic meter
m^3/d	cubic meter per day
m/h	meter per hour
m/s	meter per second
m^3/s	cubic meter per second
mg	milligram
mgd	millon gallons per day
mL	milliliter
mm	millimeter
m/s	meter per second
m^3/s	cubic meter per second
m^3/d	cubic meter per day
m^2/d	square meter per day (specific capacity; cubic meter per day discharge per meter drawdown)
N.d.	no date
N/m2	newton per square meter; pascal
ntu	nephelometric turbidity unit
NWS	National Weather Service
p.	page
PAC	Project Advisory Committee
P.E.	Professional Engineer
Ph.D.	Doctor of Philosophy
pp.	pages
ppm	parts per million
P	piezometer
Proc.	proceedings
psi	pounds per square inch
Q	flow
R	hydraulic radius
R^2	regression coefficient
RBF	riverbank filtration
s	stream slope
SI	system International [International System of Units]
τ_0	stream bed boundary shear stress
TOC	total organic carbon
U. of L.	University of Louisville
U.S.	United States
U.S.A.	United States of America
USACE	United States Army Core of Engineers

U.S E.P.A	United Stated Environmental Protection Agency
USGS	United States Geological Survey

IWA Publishing's authorised EU representative for General Product
Safety Regulations is Diane D'Arras, 15 rue Duret, 75116 Paris,
France, e-mail: safety@iwap.co.uk.

Printed and bound by CPI Group (UK) Ltd, Croydon, CR0 4YY
17/09/2026
02218802-0014